SOIL BIOTECHNOLOGY
MICROBIOLOGICAL FACTORS IN CROP PRODUCTIVITY

TO MARY

Soil Biotechnology

MICROBIOLOGICAL FACTORS IN CROP PRODUCTIVITY

J. M. LYNCH
BTech, PhD, CChem
MRSC, MIBiol

BLACKWELL SCIENTIFIC PUBLICATIONS
OXFORD LONDON
EDINBURGH BOSTON MELBOURNE

The colour plates in this book are sponsored by
ICI Plant Protection Division plc and Monsanto Europe SA

© 1983 by
Blackwell Scientific Publications
Editorial offices:
Osney Mead, Oxford, OX2 0EL
8 John Street, London, WC1N 2ES
9 Forrest Road, Edinburgh, EH1 2QH
52 Beacon Street, Boston
    Massachusetts 02108, USA
99 Barry Street, Carlton
    Victoria 3053, Australia

First published 1983

Printed and bound by
Butler & Tanner Ltd,
Frome and London

DISTRIBUTORS

USA
    Blackwell Mosby Book Distributors
    11830 Westline Industrial Drive
    St Louis, Missouri 63141

Canada
    Blackwell Mosby Book Distributors
    120 Melford Drive, Scarborough
    Ontario, M1B 2X4

Australia
    Blackwell Scientific Book Distributors
    31 Advantage Road, Highett
    Victoria 3190

British Library
Cataloguing in Publication Data

Lynch, J. M.
    Soil biotechnology.
    1. Soils   2. Plant-soil relationships
    I. Title
    631.4'1   S596.7

ISBN 0 632 00952 7

CONTENTS

PREFACE

In my previous texts, I have principally been concerned with the ecology of micro-organisms. In the present book, I will extend some of the concepts relevant to the study of soil micro-organisms and attempt to demonstrate how microbial activity might be manipulated to improve agricultural and horticultural productivity. The book is based on a course of eight lectures given to agriculture and forestry students in the University of Oxford. I hope that undergraduate and postgraduate students studying microbiology, soil science. plant science, agriculture, forestry and horticulture, as well as some farmers and growers will find it interesting.

J. M. LYNCH

Wantage
August 1982

ACKNOWLEDGMENTS

Four institutions have provided support to enable me to write this monograph. First, the Letcombe Laboratory of the Agricultural Research Council, where, for the past 11 years, I have been able to investigate some of the concepts described. In particular, I have had unfailing practical help, especially from Stephen Harper and Lynda Panting, and much valuable discussion with all the professional staff at Letcombe. I also thank the typists for producing manuscripts from my near-illegible handwriting. For the past two years I have been able to test the reaction to my ideas by my colleagues and students at the University of Oxford (as visiting lecturer), the University of Reading (as honorary member of the academic staff) and Washington State University (as visiting professor). At WSU, Lloyd Elliott and Jim Cook are sources of great inspiration to me. Just as I complete the text, I have been appointed Head of the new Department of Plant Pathology and Microbiology at the Glasshouse Crops Research Institute in Littlehampton, Sussex; I see this post as a great challenge to extend some of my ideas to protected crops.

It has been rewarding to compare and contrast field problems in Britain with those in North America, but in countries such as Poland and Mexico, which I have been fortunate to visit in the past year, the scope for the application of the ideas is equally great. Case histories have also been fascinating to investigate and I am grateful to the Monsanto Company, the ICI Plant Protection plc and the Interox Company for supporting some of my studies.

I particularly thank Mike Greaves (ARC Weed Research Organization), who critically read the entire manuscript, and John Beringer (Rothamsted), Lloyd Elliott (WSU) and John Whipps (Letcombe) who read chapters. However, any errors or omissions which remain are my responsibility. At Blackwell Scientific Publications, Robert Campbell has always been constructive in focusing the direction of my writing, and I thank Victoria Cullen for careful sub-editing.

The following generously provided me with photographs for figures: D. A. Barber (5.22), J. Bebb (13.6, 10.10), R. Campbell (2.2, 2.3, 2.4, 5.9, 5.12, 5.14, 6.14, 10.4, 10.5), M. Chandler (6.4), C. Chamswarng (6.17), S. J. Chapman (3.3), D. L. Coplin (6.16), Macaulay Institute for Soil Research and J. F. Darbyshire (3.5), M. C. Drew (5.9, 5.12, 5.14), L. F. Elliott (5.10, 5.11, 9.2, 9.3, 9.11), J. L. Faul (6.14), M. P. Greaves (8.2), E. J. Gussin (9.9, 9.10), E. Henness (2.11), ICI Plant Protection Division (6.12, 6.13), D. A. Inglis (5.3), B. G. Johnen (5.7, 5.8), M. Katouli (6.15), G. Kilbertus (2.8, 2.9, 2.10), J. Kloepper (10.6), W-H. Ko (10.3), R. Marchant (6.15), Ministry of Agriculture, Fisheries and Food (6.12), D. A. Perry (9.13, 9.14), R. T. Plumb (3.2), R. D. Prew (6.13), F. E. Sanders (6.8), J. A. Sargent (8.2, 10.10), E. L. Schmidt (6.2), G. W. F. Sewell

(9.12), J. Sitton (6.17), M. Sladdin (10.1), W. D. P. Stewart (3.4), D. A. Veal (10.9), L. H. Wullstein (5.15, 5.16, 5.17), S. F. Young (2.11, 9.9, 9.10). Gordon Lethbridge (Macaulay), David Lewis and David Read (Sheffield University) and Bernard Tinker and David Stribley (Rothamsted) provided me with valuable information in advance of publication.

The following publishers and authors allowed me to reproduce figures and tables: Fig. 2.17 (W. H. Freeman/R. J. Cook), 2.18 (Chapman & Hall/D. M. Griffin), 4.4 (R. J. Dowdell), 4.7 (Pergamon Press/S. J. Chapman), 4.9 (E. A. Paul), 4.10 (Academic Press/E. A. Paul), 6.9 (*New Phytologist*/P. B. Tinker), 8.4 (*Statens Planteavlsforsøg*/F. Eiland), 9.4 (Soil Science Society of America/L. F. Elliott), 9.7 (Pergamon Press/L. F. Elliott) and 10.2 (American Phytopathological Society/R. J. Cook); Table 2.2 (Ellis Horwood/G. Stotzky), 2.3 (Ellis Horwood/R. G. Burns), 3.2 (Academic Press/J. S. Skujins), 8.4 (Soil Science Society of America/L. F. Elliott) and 10.4 (H. Bolton).

As all authors know, writing a book takes long hours and produces many strains and frustrations. Without the support and understanding of my wife, Mary, and my children, Luke and Dominic, the effort would not have seemed worthwhile.

J. M. Lynch

When we consider the extreme antiquity, the complex character, the perpetual renovation, the constant evolution of gaseous matter and the additions of vegetable matter from growing plants which are illustrated in every fertile soil, no one can fail to be struck with wonder and admiration. The soil is a graveyard in the widest sense, and yet it is the very mainspring of new life. If any object can demonstrate the possibility of resurrection, it is the soil upon which we walk. Equally wonderful is that it is not an inert mass, but the scene of endless preparations and changes not only brought about by chemical action but by living organisms.

J. Wrightson & J. C. Newsham *Agriculture*, 2nd edn. Crosby Lockwood & Son, London, 1919.

In agriculture, horticulture and forestry, many chemical, physical and biological factors contribute to crop productivity. If seedling establishment, crop growth and final yield are poor, nutrient deficiency or adverse environmental factors can often be invoked. When all obvious options for the explanation have been considered and eliminated, the farmer, grower or forester will often then conclude that it must be a 'bug' problem. By this he or she usually means a microbiological problem because insect infestations will probably have been obvious by then. In the past this phenomenon has commonly been referred to as 'soil sickness' or the like. It is a much rarer situation for micro-organisms to be invoked as causative agents when abnormally good yields are obtained, yet micro-organisms can have as much potential to stimulate plant growth as to inhibit it. Thus, they are in part responsible for 'soil health'.

What then is the place of the microbiologist in relation to other plant and soil scientists? Pioneers such as Winogradsky and Waksman provided the foundations of the subject but in the post-pioneer period, the subject lost fashion, the microbiologist commonly only being trained in the study of pure cultures in the laboratory. Recently, however, the study of microbial ecology has again become fashionable, as witnessed by recent texts [2,4,5,7,9,13,14]. This is much welcomed because it equips the graduate microbiologist with a much better understanding of the behaviour and significance of micro-organisms in natural environments. For the microbiologist specifically interested in soils and plants, texts are available [1,6,10,17]. Whereas these texts all provide valuable knowledge, there has seldom been much consideration of how this knowledge can be applied practically in a research or advisory capacity. This is a principal objective of the present book for the student of microbiology;

however, the author emphasizes that it is not a general textbook on soil science.

'Soil fertility' refers to factors, particularly nutrients, in the soil, both biological and abiological, which lead to satisfactory plant growth. 'Soil health' was originally chosen as the title of this book, in recognition of both the positive and negative values of soil micro-organisms. So why now 'soil biotechnology' and what does this mean? The term biotechnology has recently become fashionable and has stirred public imagination but that is not the only reason for its use here. To some, genetic engineering is the major component and indeed application of fundamental knowledge from the fields of chemistry, genetics and microbiology can be expected to lead to applications in agriculture in the near future. However, it is not the only component and it is perhaps more appropriate to consider biotechnology as symbolizing the unity and coherence of a field covering many activities; it also suggests a common logic underlying bioprocesses and techniques.

In September 1981 the European Federation of Biotechnology produced the following definition, 'Biotechnology is the integrated use of biochemistry, microbiology and engineering sciences in order to achieve the technological application of the capacities of micro-organisms, cultured tissue cells, and parts thereof'. Clearly, this is relevant to the study and manipulation of soil micro-organisms in the laboratory for subsequent introduction to soils and plants but it is less appropriate when applied to the direct manipulation of microbial activities in the soil. Yet the underlying sentiment, especially the interdisciplinary and productivity attitudes, is appealing and provides some new dimensions and potentials to couple with the study of soil microbial ecology. Therefore, a definition of *soil biotechnology* as *the study and manipulation of soil micro-organisms and their metabolic processes to optimize crop productivity* is proposed. Whereas microbiology and plant pathology are the principal disciplines involved, the subject depends on several other disciplines related to soils and plants: chemistry, physics, biochemistry, genetics, plant physiology and agronomy. This definition thus avoids the argument that the first person to set a seed in the ground and produce a plant was a soil biotechnologist who did not have the wit to consider the potential positive and negative influences of micro-organisms on that process!

Students specializing in agriculture, forestry or plant and soil sciences often find the microbiology content of their courses quite small. Even with a short course, it is vital that the language of the microbiologist be understood so that subsequent communication will be easier and a freer exchange between disciplines can take place.

Some students may even be tempted to undertake a more detailed study of microbiology subsequently and indeed a basic training in plants and soils is a sound foundation for this. Although Chapters 3 and 4 introduce some basic concepts, this book is not a textbook of basic microbiology for which many texts are already available [3,12,16].

Griffiths [11] has analysed the economics of crop productivity as follows, 'Crop production is essentially an exercise in the management of *artificial* plant ecosystems. The objective is to achieve, subject to commercial constraints, the maximum yield of valuable produce per unit of land ... As long as crop production practices remain profitable there is no incentive to change them'. However, recent developments in soil microbiology, which have economic implications, may provide the necessary incentive referred to. People, for example, have for a long time recognized the obvious potential in the study of the nitrogen fixers, which could lead to a reduction in our dependence on the Haber process. However, the agrochemical industry has also been concerned about the effects of its products on soil micro-organisms (Chapter 8) and, less obviously perhaps, the side-effects which can alter the relative populations present.

The book is fairly short so that it can be purchased by students, yet the subject deserves a much greater coverage in order to present fairly the wealth of information which has accumulated. There is clearly a need for precise methodology in this subject but only a few examples are given here because of the space limitation and the reader is referred to the more detailed texts [4,8,15]. An effort has been made to present the subject in a conceptual manner and references are given to sources of more specific information. It becomes increasingly clear to the author that soil microbiology should not be separated from plant pathology because both are concerned with micro-organisms in the terrestrial environment. Much could be gained from greater cross-fertilization between these two related disciplines. A brief discussion on root pathology has been included therefore, so as to present some of the pathologist's concepts to the microbiologist (Chapter 6). Similarly, some of the concepts of soil scientists (Chapter 2) and the root physiologists (Chapter 7) are given. The latter are, of course, particularly relevant to the study of the rhizosphere (Chapter 5) and symbioses (Chapter 6). Although the overall coverage is incomplete, it is hoped that some of the gaps in knowledge and some of the recent exciting developments, to which those controlling research, development and extension services in agriculture might be inclined to divert funding in the future, have been identified.

The literature which is cited mainly relates to agriculture and horticulture but many of the problems are also relevant to forestry. Many of the practical examples are those in which the author has research experience and he hopes that the conclusions are not too biased.

The author has gained much in understanding the problems of soil microbiology by talking to farmers. It is hoped that this book will have some positive influence on their farming systems.

References

1 Alexander M. (1977) *An Introduction to Soil Microbiology*, 2nd edn. John Wiley & Sons, New York.
2 Atlas R. M. & Bartha R. (1981) *Microbial Ecology: Fundamentals and Applications*. Addison-Wesley, Reading, Massachusetts.
3 Brock T. D. (1979) *Biology of Micro-organisms*, 3rd edn. Prentice-Hall, Englewood Cliffs, New Jersey.
4 Burns R. G. & Slater J. H. (eds) (1982) *Experimental Microbial Ecology*. Blackwell Scientific Publications, Oxford.
5 Campbell R. E. (1977) *Microbial Ecology*. Blackwell Scientific Publications, Oxford.
6 Dommergues Y. R. & Krupa S. V. (eds) (1978) *Interactions between Non-pathogenic Soil Micro-organisms and Plants*. Elsevier, Amsterdam.
7 Ellwood D. C., Hedger J. N., Latham M. J., Lynch J. M. & Slater J. H. (eds) (1980) *Contemporary Microbial Ecology*. Academic Press, London.
8 Grainger J. M. & Lynch J. M. (eds) (1983) *Microbiological Methods for Environmental Biotechnology*. Academic Press, London.
9 Grant W. D. & Long P. E. (1981) *Environmental Microbiology*. Blackie & Son, Glasgow.
10 Gray T. R. G. & Williams S. T. (1971) *Soil Micro-organisms*. Oliver & Boyd, Edinburgh.
11 Griffiths E. (1981) Iatrogenic plant diseases. *Annual Review of Phytopathology*, **19**, 69–82.
12 Hawker L. E. & Linton A. H. (eds) (1979) *Micro-organisms: Function Form and Environment*, 2nd edn. Edward Arnold, London.
13 Loutit M. & Miles J. A. R. (eds) (1978) *Microbial Ecology*. Springer-Verlag, Berlin.
14 Lynch J. M. & Poole N. J. (eds) (1979) *Microbial Ecology: A Conceptual Approach*. Blackwell Scientific Publications, Oxford.
15 Parkinson D., Gray T. R. G. & Williams S. T. (1971) *Methods for Studying the Ecology of Soil Micro-organisms*. Blackwell Scientific Publications, Oxford.
16 Stanier R. Y., Adelberg E. A. & Ingraham J. L. (1976) *General Microbiology*, 4th edn. Collier Macmillan, London.
17 Walker N. (ed.) (1975) *Soil Microbiology. A Critical Review*. Butterworth, London.

THE SOIL AS A HABITAT FOR MICRO-ORGANISMS

I sing once more
The mild continuous epic of the soil,
Haysel and harvest, tilth and husbandry;
I tell of marl and dung, and of the means
That break the unkindly spirit of the clay.

V. Sackville-West The Land. (1926) In *Collected Poems*, vol. I. The Hogarth Press, London, 1933.

In order to attempt to understand the physiology of a micro-organism in soil, it is essential that the physical and chemical nature of the soil environment be understood. Only a few salient features and concepts are mentioned here and the reader is referred to more detailed introductions to soil science [7,25,28].

Generally, the soil biomass and most individual groups of micro-organisms decrease with increasing depth down the soil profile. There are, of course, exceptions with certain organisms and in some soils, e.g. in peat and forest soils where the surface litter can generate acidity, so increasing the population of acid-tolerant organisms. However, the general profile distribution is not surprising because available energy substrates and inorganic nutrients are present in greatest amounts near the surface. It is, therefore, perhaps more relevant to consider the other physical and chemical factors which govern the distribution of micro-organisms in soil.

2.1 Factors of ecological significance

In any laboratory culture system or natural environment, physical and chemical factors govern microbial growth and activity. In natural environments, particularly soil, far more factors have to be considered. Physical factors include temperature but less obvious are osmotic pressure, surface tension, viscosity, radiation (visible, ultra-violet and ionizing), adsorption phenomena and spatial constraints. In the chemical environment of soils it is necessary to consider water activity, pH, the quality and quantity of organic and inorganic nutrients and gases, growth-promoting and growth-inhibiting substances, and oxidation–reduction potentials. Each microbial species will have an optimum for each physical or chemical factor and its growth, or activity, will decline either side of that optimum, governing its contribution to the total population. Thus, in Fig. 2.1 at 1.5, A would dominate, at 3·4, each would balance, and at 5, B would dominate. However, as there are many factors, the overall dominance depends on an integration of the responses to all factors.

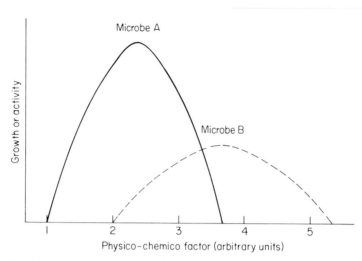

Fig. 2.1   Response of micro-organisms to factors in the soil environment.

Soil structure

The principal inorganic components of soil are sand, silt and clay.
The sizes of those particles relative to micro-organisms and plant
roots are given in Table 2.1. Figure 2.2 shows a fungus crossing a
sand grain. Clay particles are often a similar size to the bacterial cell
but they can also be much smaller. Interactions between clays and
microbial cells *in vitro* are illustrated in Figs 2.3 and 2.4. Like bacteria, clays carry a net electronegative charge. The principal clays of
soil are kaolinite, illite and montmorillonite which carry charges of
about 5, 20 and 100 milliequivalents $100\,g^{-1}$ respectively. The charge
originates from the Gouy–Chapman diffuse electrical double layer
(Fig. 2.5) [6]. Once a surface has become charged it will attract
counter (oppositely charged) ions from the surrounding aqueous

Table 2.1   Sizes of soil constituents.

	Diameter or thickness (μm)
Inorganic constituents	
Sand	50–2000
Silt	2–50
Clay	<2
Micro-organisms	
Bacteria	0·5–1·0
Actinomycetes	1·0–1·5
Fungi	0·3–10
Plants	
Root hairs	10–14
Root cylinders	40–100

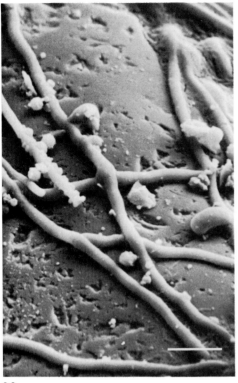

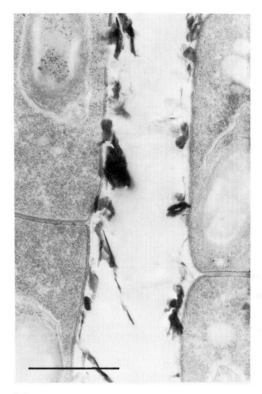

2.2

2.3

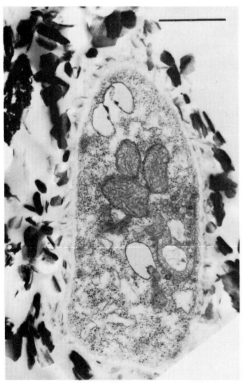

Fig. 2.2 Frozen hydrated specimen of
Gaeumannomyces graminis crossing a sand grain.
There is almost no water film around most hyphae
except at the top right. Bar marker = 10 μm.
(Photograph by R. Campbell, Bristol University.)

Fig. 2.3 *Bacillus mycoides* growing in a liquid
culture with kaolin to show the attachment of the
clay to its surface. Bar marker = 1 μm. (Photograph
by R. Campbell, Bristol University.)

Fig. 2.4 *Gaeumannomyces graminis* growing in a
liquid culture with kaolin to show the attachment of
the clay to its surface. Bar marker = 1 μm.
(Photograph by R. Campbell, Bristol University.)

2.4

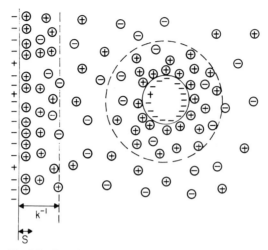

Fig. 2.5 Electrical double layer associated with a planar surface and a spherical particle. S = the thickness of a layer of adsorbed counter ions referred to as the stern layer; k^{-1} = the thickness of the diffuse double layer.

phase. This process is opposed by the thermal motion of the counter ions, tending to distribute them evenly throughout the aqueous phase. The effects of electrostatic attraction and thermal motion on the counter ions lead to the formation of a region next to the charged surface where the concentration of the counter ions is greater than in the rest of the aqueous phase. Thus, bacteria can only interact with clays because the charges on the cell and the clay become polarized or, alternatively, they are 'bridged' by a metal ion (Fig. 2.6). However, when clays attach to micro-organisms, they can have many and varied influences on their activities (Table 2.2).

Organic matter, particularly polysaccharides, which originates from cells, and the native soil organic matter, such as humic acids (Chapter 4), surrounds the particles and micro-organisms. These materials provide sources of adhesive forces which may also have a charged nature (more detailed accounts of particle–micro-organism interactions are available in monographs [3,22]). Thus, micro-organisms are probably major determinants of soil structure (Fig. 2.7). Rhizosphere bacteria are particularly important because very stable soil structures are formed under grass where there is a much greater rhizosphere biomass than that in arable soils.

Multilayered clay particles also interact with enzymes and substrates. Urease is incorporated into organic-matter molecules and is absorbed on the outside of clays. Urea, being a small molecule, can rapidly diffuse to these sites but the protein cannot [4]. Further examples of these interactions are given in Table 2.3.

The size of pores and their continuity is particularly important in

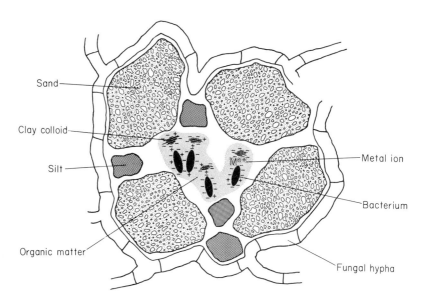

Fig. 2.6 Hypothetical spatial arrangement of micro-organisms in relation to soil particles.

soil aeration. Soil aggregates (Fig. 2.8), viewed by transmission electron microscopy, have some pores which are closed (Fig. 2.9) and others which are open (Fig. 2.10) to allow the diffusion of gases [19]. The presence of water also governs the degree of continuity in the pore space; as the suction or tension on the soil is increased, more gas space is created.

The way in which soil particles come together to form aggregates, clods or crumbs gives soil its structure [1,18]. Figure 2.11 shows bacteria adhering to particles within an aggregate. Tilth is dependent on the size distribution of particles and their mellowness or rawness. Good soil structures hold sufficient water to prevent moisture deficits around plant roots during dry periods but also allow sufficient drainage to prevent waterlogging during wet periods. Structure is also important at seeding; the aggregates in the seed-bed must be sufficiently large to prevent wind erosion but must not be so large as to provide a mechanical barrier to the germinating seed. Figure 2.12 shows how capping of an unstable soil has retarded seedling establishment. Such structural qualities are difficult to measure physically but water stability is a useful criterion of good structure. Even this can be measured in several ways. Many investigators of the effects of micro-organisms on the water stability of soils have shaken soil in a wet state on a series of sieves: soils with the greatest stability are least prone to slaking and stay on the coarser mesh [8]. Others have

Table 2.2 Some effects of montmorillonite (M) on micro-organisms at different levels of experimental complexity. (CEC = the cation exchange capacity.) (From Stotzky [26].)

Level of experimental complexity	Effect
Field observations	*Fusarium* wilt of bananas: faster spread in soils without M
	Histoplasma capsulatum: isolated essentially only from soils without M
	Coccidiodes immitis: isolated from both soils with and without M
	Other mycotic pathogens of humans: so far isolated only from soils without M
	Enzootic leptospirosis: so far isolated essentially only where soils and sediments do not contain M
Pure culture studies	Fungi:
	respiration, radial growth and spore germination decreased by concentrations greater than 2% (w/v) (related to increases in viscosity which, in turn, reduces access to O_2)
	Bacteria:
	respiration stimulated
	stimulation due to buffering effect (pH)
	stimulation related to CEC and not particle size or specific surface of clays
	lag phase of growth decreased
	protected against hypertonic osmotic pressures (related to CEC)
	Protect both bacteria and fungi against heavy metal toxicities (related to CEC)
	Sorb organic volatiles from germinating seeds and bacteria
Spread through soil (soil replica plate studies)	Fungi slower in presence of M; bacteria faster in presence of M
	Competitive effects of bacteria against fungi greater in presence of M: related to CEC, 'buffering effect' and more rapid utilization of nutrients by bacteria (competition)
	Conjugation of bacteria stimulated by presence of M
	Protects bacteria and fungi against heavy metal toxicities (related to CEC)
Metabolism in soil (autotrophic and heterotrophic)	Nitrification rate enhanced by M
	Decomposition of aldehydes enhanced by M
	Decomposition of other organics either enhanced, decreased or unaffected by M
	Protects against inhibition by SO_2 of nitrification
	Protects against inhibition by heavy metals of organic matter decomposition
Greenhouse studies	Spread of fusarial wilts decreased by presence of M in soil
Field studies	Incorporation of M into soils: effect on persistence and spread of plant and animal pathogens (to be conducted)

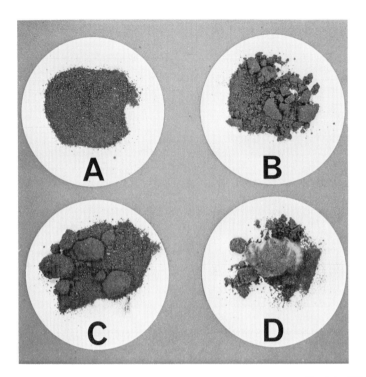

Fig. 2.7   Effect of washed suspensions of soil micro-organisms on the aggregation of sandy soil (Newport series) after oven drying at 55° C for six hours. (A) sterile; (B) mixed population; (C) *Azotobacter chroococcum* (which produces extracellular polysaccharide); (D) *Mucor hiemalis*.

Fig. 2.8   Section through an aggregate showing the evidence for the distribution of bacteria, the presence of type 1 (closed) and type 2 (open) pores with air space between organic (MO) and inorganic (MI) materials, clay particles (A) and fungal residue (C). The unlabelled arrow probably points to a type 1 pore. (Photograph by G. Kilbertus [19], Nancy University.)

Table 2.3 Influence of soil colloids on substrate decay and microbial growth. (From Burns [4].)

Colloid surface phenomenon	Effect on substrate decay and/or microbial growth (relative to that in the absence of clay/humic colloid)
Juxtaposes microbe (or enzyme) and substrate	Stimulation
Orients enzyme beneficially relative to substrate	Stimulation
Functions as buffer during metabolism	Stimulation
Adsorbs inhibitory metabolites	Stimulation
Retains H_2O film = prevents desiccation	Stimulation
Concentrates inorganic nutrients	Stimulation
Supplies inorganic micronutriénts (clay)	Stimulation
Protects microbes from predators	Stimulation
Inactivates phages	Stimulation
Produces soluble substrate (humic enzyme complex)	Stimulation
Adjusts carbon/nitrogen ratios (humic)	Stimulation
Allows cometabolism of adsorbate (humic)	Stimulation
Performs abiological decay in a biological sequence	Stimulation
Absorbs microbe (or enzyme) distant from substrate	Inhibition
Intercalates substrate = inaccessible to microbe (clay)	Inhibition
Incorporates substrate into humic polymer = inherited resistance	Inhibition
Inactivates enzymes due to structural changes	Inhibition
Masks active site of enzyme	Inhibition
Increases viscosity = retards O_2 diffusion	Inhibition
Entraps microbe in colloidal aggregate = limited O_2, nutrients, etc.	Inhibition

measured the number of drops of water from a burette needed to destroy an aggregate [13] or have shaken the soil in water and measured the slaking by the turbidity of the suspensions produced [9,20]. The details of each method have been selected somewhat arbitrarily but generally it has been shown that the water stability of soils increases with the added biomass. However, in one investigation, this was only true in a clay soil and not in a more weakly structured silt loam. In the latter, bacteria and a yeast improved water stability but a common soil fungus, *Mucor heimalis*, decreased it [20] (Fig. 2.13). The mechanism is uncertain for, as there was no incubation period, the fungus would not have destroyed the natural aggregating agents which might occur normally under field conditions. It is possible that the fungal hyphae prevented the natural aggregation by the small natural population of bacteria present (Fig. 2.14). Since the substantial interest in the contribution of microorganisms to soil structure about 30 years ago, there have been

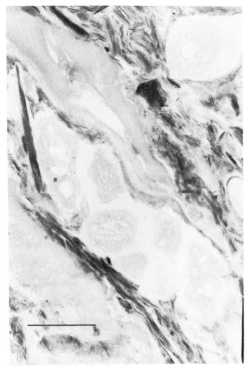

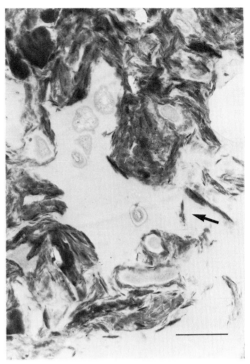

2.9

2.10

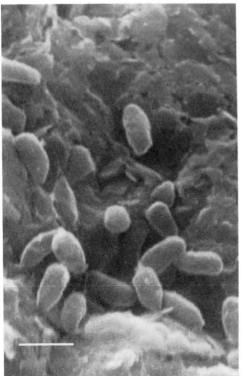

Fig. 2.9 Type 1 (closed) pore containing bacteria. Bar marker = 1 μm. (Photograph by G. Kilbertus [19], Nancy University.)

Fig. 2.10 Type 2 pore containing bacteria with the opening arrowed. Bar marker = 1 μm. (Photograph by G. Kilbertus [19], Nancy University.)

Fig. 2.11 *Pseudomonas* sp. within an aggregate of a clay soil (Denchworth series). Bar marker = 1 μm. (Photograph by E. Henness and S. F. Young, Letcombe Laboratory.)

2.11

Fig. 2.12 Capping of a silt loam soil in response to irrigation. The clay loam soil is water stable and seedlings have emerged more rapidly.

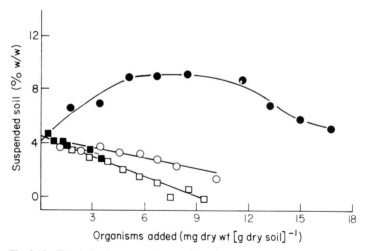

Fig. 2.13 Effect of micro-organisms on the stability of a silt loam (Hamble series). Solid circle = *Mucor hiemalis*; open circle = *Azotobacter chroococcum*; solid square = *Lipomyces starkeyi*; open square = *Pseudomonas* sp. (From Lynch [20].)

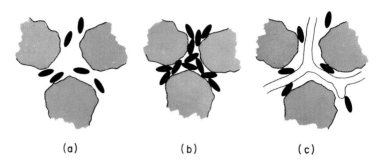

Fig. 2.14   Possible interpretation of data shown in Fig. 2.11. (a) A natural population of bacteria aggregating soil particles. (b) An increased bacterial biomass promoting cementation. (c) Fungal hypha preventing the cementing action of the natural population.

relatively few studies. The importance of soil structure to crop production justifies more studies using modern techniques.

2.3   Soil aeration

A further attribute of a good soil structure is that there should be sufficient gas-filled pores to allow gas exchange with the atmosphere and reduce the chances of localized anaerobic pockets or microsites developing which can have many and varied effects on the growth of plant roots. These effects include the direct ones of anoxia on plant physiological processes, modification of endogenous metabolism within the roots, including changes in the accumulation of growth regulators, and changes in the pathogen and saprophyte populations which can produce phytotoxic substances around roots [5]. To reduce the risk of anaerobic pockets developing, there should be a continuity of the pore space within soil. When the pores become blocked and active microbial metabolism around roots or decomposing plant residues takes place in the sealed pockets, oxygen becomes depleted (Fig. 2.15). Similar, smaller pockets probably also develop within soil aggregates where organic matter is decomposed and when the concentration of oxygen falls to about $3 \times 10^{-6}\,\mathrm{mol}^{-1}$ (less than $1\%\,\mathrm{v/v}$) there appears to be a switch from aerobic to anaerobic metabolism [10], although this concept is difficult to prove. However, the important aspect is that all soils have a heterogenous distribution of oxygen; so-called 'anaerobic soils' have pockets of air and 'aerated soils' have anaerobic pockets.

Micro-organisms generally have much greater specific rates of oxygen uptake than plant roots or seeds (Table 2.4) and hence there is potential for oxygen competition to take place when the supply of oxygen is restricted. This has never been demonstrated for roots

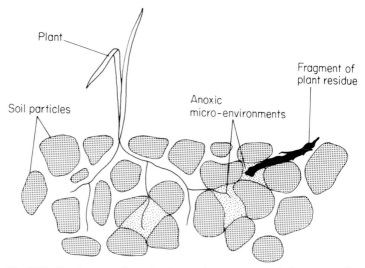

Fig. 2.15 Development of anoxic micro-environments around plant roots and residues when the voids between soil particles are sealed.

Table 2.4 Specific oxygen uptake rates (qO_2) for some micro-organisms, roots and seeds.

	qO_2 ($\mu l\ mg^{-1}\ dry\ wt.h^{-1}$)
Azotobacter sp.	5000
Escherichia coli	200
Mucor hiemalis	150
Penicillium sp.	20
Sterile cereal roots	4
Germinating barley seed	0·25

although Griffin [11] has outlined how a model could be produced for this. Using a Gilson microrespirometer and experiments in different gaseous mixtures, bacterial seed [15] and fungal seed [16] competition for oxygen has been demonstrated and this can be a cause of poor seed germination in the field (Chapter 9).

When all the oxygen dissolved in the soil water has been used in respiration and chemical oxidations, anaerobic micro-organisms respire via a sequence of components acting as alternative electron acceptors to oxygen. Thus, as a consequence of microbial respiration, the redox potential (E_h) of a soil declines. The more highly reduced a compound is, the greater the amount of energy released when it is oxidized:

$$\underset{\substack{\text{Reduced}\\\text{state}}}{AH_2} + B \rightarrow \underset{\substack{\text{Oxidized}\\\text{state}}}{A} + BH_2.$$

The E_h is the tendency for energy to be released in an oxidation reaction. More precisely, it is the tendency for the reduced compound to release electrons and become oxidized. The more negative system is oxidized (losing electrons) while the more positive system is reduced (gaining electrons). Such reactions apply to chemical reactions in soils and to the biochemical pathways in the microbial cell. The electron transport chain of the cell (the stepwise transfer to electrons) is able to convert reducing equivalents, such as reduced pyridine nucleotides or flavoproteins, to energy as ATP. The terminal electron acceptors can range from O_2 to NO_3^-, NO_2^-, SO_4^{2-} or CO_2. Some redox changes with the organisms responsible are given in Table 2.5. The significance of these in a slurry of soil is illustrated in Fig. 2.16. Hamilton [14] provides a detailed account of electron transport within the microbial cell and Ponnamperuma [24] gives an account of the effect of redox changes in soils.

Table 2.5 Some redox couples of soil bacteria.

	Redox couple (terminal electron acceptor)	E_h (mv)
(Aerobic respiration)	$\frac{1}{2}O_2/H_2O$	+820
Bacillus, Clostridium	NO_3^-/NO_2^-	+433
Paracoccus, Thiobacillus	NO_2^-/NO	+350
Clostridium, Desulfovibrio	Fumarate/succinate	+ 33
Desulfovibrio	SO_4^{2-}/SO_3^{2-}	− 60
Methanobacterium	CO_2/CH_4	−350

The changing aeration of soils has important consequences for micro-organisms and enzymes; e.g. the effect on nitrogenase is discussed in Chapter 6. The flooding of soils for rice production is an extreme case of the waterlogged soil; the microbial ecology of these soils has been reviewed [27].

2.4 Acidity and alkalinity

The activities of all microbial enzymes, about 1000 per cell, are dependent on the H^+ ion and therefore soil pH influences these. The pH of a soil will also govern species diversity and it is perhaps surprising that there have been relatively few studies of this. The microbial group most studied in this respect is the antibiotic-producing actinomycetes. *Streptomyces*, for example, cannot grow

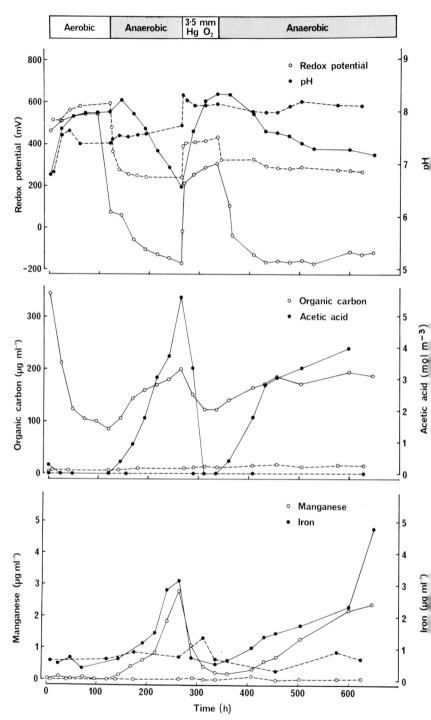

Fig. 2.16 Effect of oxygen on the physiochemical changes taking place in a stirred slurry of 300 g of a silty loam (Hamble series) with or without ground wheat straw (50 g) in distilled water (3.2 l). Solid lines = soil mixed with straw; broken lines = soil only. (From Lynch & Gunn [21].)

below pH 7·5 but its species are routinely found in acid forest soils. This is because nitrogen-rich plant residues are decomposed to give localized NH$_3$ release to increase the pH above 7·5 [29]. In wet soils, however, it should be recognized that bacterial fermentation, yielding organic acids, can cause the pH to fall. Furthermore, it has recently been shown that acidophilic streptomycetes are also involved in the decomposition process and the resulting ammonification can lead to the activity of neutrophiles [30].

2.5   Temperature

The temperature minimum of an organism is governed by the changes of enzyme proteins and by the fluidity of membranes, which in turn is governed by the degree of unsaturation of the fatty acids in the membranes. Enzyme denaturation is the factor limiting the temperature maximum and this varies greatly between soil organisms (Fig. 2.17). Clearly, organisms vary greatly in the temperature maxima, minima and optima for growth and this is a further factor influencing species diversity. Thermophiles can proliferate in tropical

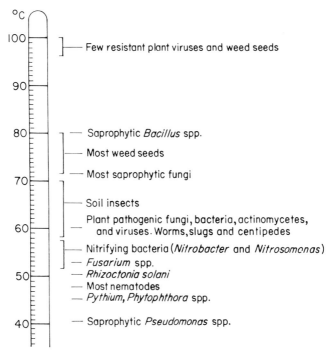

Fig. 2.17   Temperatures necessary to kill various groups of soil organisms based on a 30 min exposure to moist heat (10 min for nitrifying bacteria). (From Baker & Cook [2].)

soils and psychrophiles are well adapted in arctic soils. Activity of all organisms is sometimes expressed as a Q_{10}, the increase in rate of a process (expressed as a multiple of the initial rate) produced by raising the temperature $10°C$. This is often between 2 and 3 for biological processes but it is governed by the Arrhenius equation and is therefore similar for chemical processes.

2.6 Water

The gravimetric water content of a soil, w, is expressed as g water g^{-1} soil dried for 24 h at $105°C$, whereas the volumetric water content, θ, is expressed as cm^3 water cm^{-3} soil. The former is the easiest to determine but the latter is usually a more useful measure because it allows for shrinkage and expansion of soils which accompany changes in water content. The bulk density, ρ_b (g solids cm^{-3} total volume) is related by:

$$\theta = w\rho_b.$$

However, none of these measures indicate the water actually available to plants and micro-organisms when the water in soil is subjected to suction by them. The most useful measure for this is the water potential, which consists of matric and osmotic components. The matric potential is concerned with the retention of water in the voids between soil particles (Fig. 2.18). The potentials are governed by the surface-tension forces and are proportional to the radius of the menisci of the water, which can be analysed from the principles of capillary rise. If the pores have an 'effective' or 'equivalent' capillary radius, r, the suction necessary to absorb water, the matric potential τ (dyn cm^{-2}), can be expressed as:

$$\tau = 2\gamma r^{-1}$$

where γ is the surface tension of water (73 dyn cm^{-1}). It is clear from the figure that as suction is increased, r becomes smaller, less water is present and hence τ becomes larger. Suction can be expressed as:

$$1 \text{ bar} = 10^6 \text{ dyn cm}^{-2} = 100 \text{ kPa} = 1022 \text{ cm H}_2\text{O} = 75 \text{ cm}$$
$$\text{Hg} = 0.987 \text{ atm.}$$

The osmotic potential of water, π, results from the presence of solutes and is less important than the matric component in most soils. The solutes decrease the entropy and increase the order:

$$\pi = -RT\rho v m\varphi \times 10^{-9}$$

where R is the gas constant (8.31×10^7), T is the temperature ($°K$), v the ions per molecule, which is taken as 1 for non-ionic solutes, m is the molality, φ is the osmotic coefficient and ρ is the density.

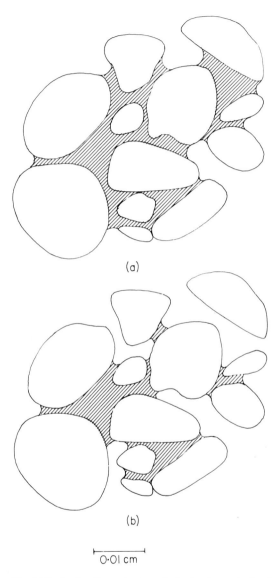

(a)

(b)

⊢———⊣
0·01 cm

Fig. 2.18 Section through a hypothetical soil sample showing the distribution of water at two matric potentials: (a) −100 cm water; and (b) −200 cm water. (From Griffin [12].)

Another expression of the water available to a micro-organism is water activity, *Aw*, a term widely used by microbiologists studying food spoilage. This is related to the equilibrium relative humidity (ERH), the humidity of the vapour in equilibrium with the soil:

$Aw = \text{ERH } 100^{-1}$.

The availability of water in soil affects the species diversity, survival, movement and activity of micro-organisms. To test the effects of water potential stress on micro-organisms, the osmotic potential is usually controlled by adjusting the solutes available to them. In general, fungi can tolerate lower potentials than bacteria which are limited to about -200 bar. Some examples are given in Table 2.6. A more detailed examination of water potential relations in soil microbiology is available [23].

Table 2.6 Tolerance of the solute-controlled water potential on micro-organisms. (After Harris [17].)

ψ bar	A_w	NaCl (w/v)	Sucrose (w/v)	Example
$-$ 15	0·99	2·0	17	*Spirillum, Rhizobium*
-100	0·93	12·3	52	*Clostridium, Mucor*
-250	0·83	25·5	70	*Micrococcus, Penicillium*
-650	0·62	—	83	*Xeromyces, Saccharomyces*

2.7 Conclusion

It is often difficult to separate the many factors of the soil environment from their effect on micro-organisms. Water controls not only soil-water potential, for example, but also the aeration and the redox potential. The contribution of each of these factors can only be assessed in isolation and this usually means autoecological studies in the laboratory under defined conditions employing chemostats (section 4.3) and other equipment, giving maximum control on the environment of the micro-organism.

References

1 Allison F. E. (1968) Soil aggregation—some facts and fallacies as seen by a microbiologist. *Soil Science*, **242**, 136–43.
2 Baker K. F. & Cook R. J. (1974) *Biological Control of Plant Pathogens*. W. H. Freeman, San Francisco.
3 Berkeley R. C. W., Lynch J. M., Melling J., Rutter P. R. & Vincent B. (eds) (1980) *Microbial Adhesion to Surfaces*. Ellis Horwood, Chichester.
4 Burns R. G. (1980) In *Microbial Adhesion to Surfaces*, eds Berkeley R. C. W., Lynch J. M., Melling J., Rutter P. R. & Vincent B., pp. 249–62. Ellis Horwood, Chichester.
5 Drew M. C. & Lynch J. M. (1980) Soil anaerobiosis, micro-organisms and root function. *Annual Review of Phytopathology*, **18**, 37–67.
6 Fletcher M., Latham M. J., Lynch J. M. & Rutter P. R. (1980) In *Microbial Adhesion to Surfaces*, eds Berkeley R. C. W., Lynch J. M., Melling J., Rutter P. R. & Vincent B., pp. 67–78. Ellis Horwood, Chichester.

7 Foth H. D. (1978) *Fundamentals of Soil Science*, 6th edn. John Wiley & Sons, Chichester.

8 Geoghegan M. J. & Brian R. C. (1948) Aggregate formation in soil. I. Influence of some bacterial polysaccharides on the binding of soil particles. *Biochemical Journal*, **43**, 5–13.

9 Gilmour C. M., Allen O. N. & Truog E. (1948) Soil aggregation as influenced by the growth of mold species, kind of soil and organic matter. *Soil Science Society of America Journal*, **13**, 292–6.

10 Greenwood D. J. & Berry G. (1962) Aerobic respiration in soil crumbs. *Nature (London)*, **195**, 161–3.

11 Griffin D. M. (1968) A theoretical study relating the concentration and diffusion of oxygen to the biology of organisms in the soil. *New Phytologist*, **67**, 561–7.

12 Griffin D. M. (1972) *Ecology of Soil Fungi*. Chapman & Hall, London.

13 Griffiths E. & Jones D. (1965) Microbiological aspects of soil structure. I. Relationships between organic amendments, microbial colonization and changes in aggregate stability. *Plant and Soil*, **23**, 17–33.

14 Hamilton W. A. (1979) Microbial energetics and metabolism. In *Microbial Ecology. A Conceptual Approach*, eds Lynch J. M. & Poole N. J., pp. 22–44. Blackwell Scientific Publications, Oxford.

15 Harper S. H. T. & Lynch J. M. (1979) Effects of *Azotobacter chroococcum* on barley seed germination and seedling development. *Journal of General Microbiology*, **112**, 45–51.

16 Harper S. H. T. & Lynch J. M. (1981) Effects of fungi on barley seed germination. *Journal of General Microbiology*, **122**, 55–60.

17 Harris R. F. (1981) Effect of water potential on microbial growth and activity. In *Water Potential Relations in Soil Microbiology*, eds Parr J. F., Gardner W. R. & Elliott L. F., pp. 23–95. Soil Science Society of America, Madison, Wisconsin.

18 Harris R. F., Chesters G. & Allen O. N. (1966) Dynamics of soil aggregation. *Advances in Agronomy*, **18**, 107–69.

19 Kilbertus G. (1980) Etude des microhabitats contenus dans les agregats du sol. Leur relation avec la biomasse bacterienne et la taille des procaryotes presents. *Revue d'Ecologie et de Biologie du Sol*, **17**, 543–7.

20 Lynch J. M. (1981) Promotion and inhibition of soil aggregate stabilization by micro-organisms. *Journal of General Microbiology*, **126**, 371–5.

21 Lynch J. M. & Gunn K. B. (1979) The use of the chemostat to study the decomposition of wheat straw in soil slurries. *Journal of Soil Science*, **29**, 551–6.

22 Marshall K. C. (1976) *Interfaces in Microbial Ecology*. Harvard University Press, Cambridge, Massachusetts.

23 Parr J. F., Gardner W. R. & Elliott L. F. (eds) (1981) *Water Potential Relations in Soil Microbiology*. Soil Science Society of America, Madison, Wisconsin.

24 Ponnamperuma F. N. (1972) The chemistry of submerged soils. *Advances in Agronomy*, **24**, 29–96.

25 Russell E. W. (1973) *Soil Conditions and Plant Growth*, 10th edn. Longman, London.

26 Stotzky G. (1980) In *Microbial Adhesion to Surfaces*, eds Berkeley R. C. W., Lynch J. M., Melling J., Rutter P. R. & Vincent B., pp. 231–47. Ellis Horwood, Chichester.

27 Watanabe I. & Furusaka C. (1980) Microbial ecology of flooded rice soils. *Advances in Microbial Ecology*, **4**, 125–68.

28 White R. E. (1979) *Introduction to the Principles and Practice of Soil Science*. Blackwell Scientific Publications, Oxford.

29 Williams S. T. & Mayfield C. I. (1971) Studies on the ecology of actinomycetes in

soil. III. The behaviour of neutrophilic streptomycetes in acid soils. *Soil Biology and Biochemistry*, **3**, 197–208.

30 Williams S. T. & Robinson C. S. (1981) The role of streptomycetes in the decomposition of chitin in acidic soils. *Journal of General Microbiology*, **127**, 55–63.

MICRO-ORGANISMS AND ENZYMES IN THE SOIL

Not only has vegetative power been supposed to be requisite for the organization of matter in animated beings, but that it might change an animal to the vegetable state, and the vegetable again to an animal; that it acts on plants while living and when dead, regenerates them in new beings. These are the animalcula of infusions, which cannot strictly be called animals, but beings simply *vital*.

Lazzaro Spallanzari *Opuscoli di Fisica Animale, e Vegetabile*, 1776. (Translation by J. G. Dalyell in *The Tracts on the Nature of Animals and Vegetables*. Creech & Constable, Edinburgh, 1799.)

The study of soil micro-organisms is usually approached at three levels: (1) the determination of their form and arrangement in the soil; (2) isolation and characterization; and (3) the detection of activity. This produces information on the types, biomass (the living material) and function of those present. Studies are conducted both of communities in the natural environment (synecology) and of single species in laboratory conditions (autecology) but to form a complete understanding of soil micro-organisms, it is necessary to combine both approaches, with special emphasis on spatial and temporal considerations.

3.1 Soil microbial population

All known types of micro-organisms occur in soils. One of the most important divisions is that between the *prokaryotes* (bacteria and blue-green algae) and the more highly organized *eukaryotes* (fungi, algae and protozoa) [21]. Distinctions are primarily at the structural and ultrastructural level [24]. Prokaryotes do not have a nuclear membrane, endoplasmic reticulum and mitochondria, whereas eukaryotes have all these structural features. Whittaker [27] proposed a five kingdom scheme to divide micro-organisms on the basis of discipline, ultrastructure and trophic modes (Fig. 3.1). The relative sizes of micro-organisms in soil can be seen from the various plates in this book and the ranges are given in Table 3.1. It would of course be quite beyond the scope of this book to list all the types of micro-organisms present in soil but the reader should be aware of the major sources of information. Bacteria and actinomycetes are classified in Bergey's manual [5] and a shorter edition is also available [10]. Soil fungi are also described in major [8] and shorter [9] volumes.

Fungi are the major contributors to the soil biomass and can account for about 70 per cent by weight [1]. Bacteria can in some circumstances become more important in the rhizosphere (Chapter

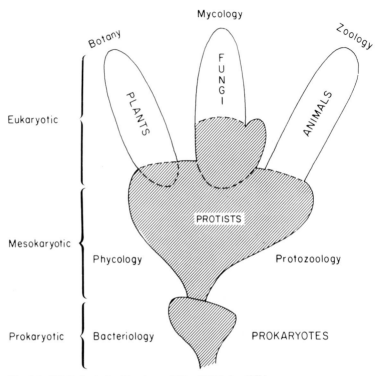

Fig. 3.1 Whittaker's five kingdoms. (After Whittaker [27].)

Table 3.1   Relative sizes of micro-organisms and proteins.

	Size (μm)
Proteins	0·001–0·05
Bacterial viruses (phage)	0·05–0·10 (head), 0·2 (tail)
Plant viruses	0·02–0·3
Bacteria	0·5–2 × 1–8
Actinomycetes	0·5–2 (diameter)
Cyanophyceae	2–5 (diameter)
Algae	3–50 (diameter)
Fungi	3–50 (diameter)
Protozoa	14–600

5). In flooded soils, anaerobic bacteria increase in importance. Blue-green algae are photosynthetic and are agriculturally important where there is an abundant source of water on the soil surface and sunlight as they fix atmospheric nitrogen.

Viruses in soil are usually associated with clay particles (section 2.2). Viruses usually only enter plant roots through lesions or cracks which formed for some other reason; alternatively they are carried on insects, fungi and sometimes bacteria as vectors. Figure 3.2 shows red clover necrotic mosaic virus (RCNMV) which is soil borne. This gives an idea of virus size but the forms of the various viruses differ greatly, with the symmetry being helical, cubical or binal. Viruses contain a nucleic-acid core with a protein coat. Some are naked but others are enveloped. They contain one or two strands of DNA or RNA. Plant viruses are grouped largely on the basis of the host cells that they infect, whereas animal and bacterial viruses are classified at the family level.

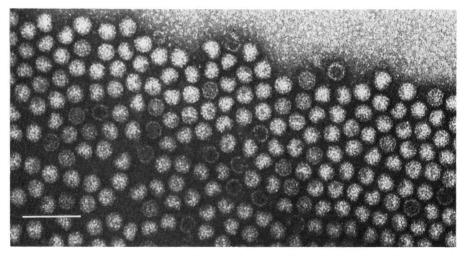

Fig. 3.2   Purified particles of red clover necrotic mosaic virus. Bar marker = 0·1 μm. (Photograph by R. T. Plumb, Rothamsted Experimental Station.)

Winogradsky [28] provided a useful basis for distinguishing bacterial groups as *autochthonous* (where there is a low but steady level of activity on native soil organic matter) and *zymogenous* (where there is rapid metabolism of freshly available organic matter). *Arthrobacter* spp., which are rods in the early vegetative state and later form cocci, cystites and buds (Fig. 3.3), survive for long periods in the bulk soil without a substantial substrate input and are good examples of the former group, whereas *Pseudomonas* spp. are very common around roots where there is abundant supply of readily assimilable carbon and these characterize the zymogenous microflora. However, it is possible that these bacteria will have some autochthonous and zymogenous characteristics and it seems unwise to apply the definitions too rigorously.

Other important distinguishing features of micro-organisms are

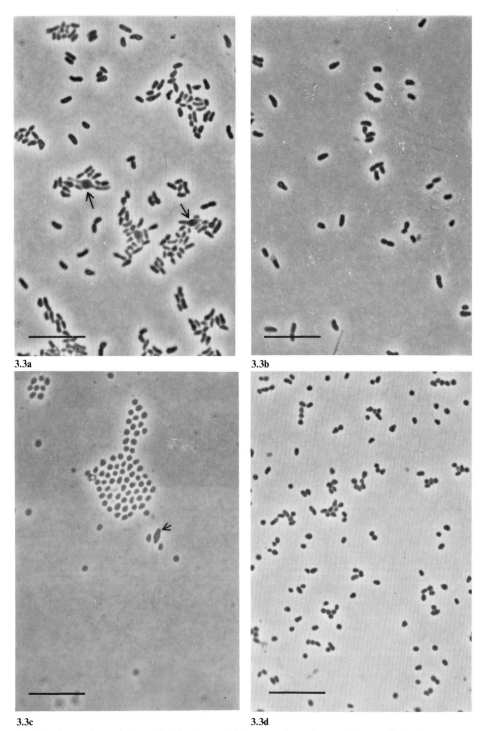

3.3a

3.3b

3.3c

3.3d

Fig. 3.3 Size and morphology of *Arthrobacter globiformis* under carbon and nitrogen limitation at a range of dilution rates. D is the dilution (growth) rate. (a) Nitrogen-limited rods (D = 0·35 h^{-1}). The arrows show enlarged cells with germinating rods. (b) Carbon-limited rods (D = 0·35 h^{-1}). (c) Nitrogen-limited cocci/rods (D = 0·1 h^{-1}). The arrow shows an enlarged cell. (d) Carbon-limited cocci/rods (D = 0·1 h^{-1}).

Bar markers = 10 μm. (Photograph by S. J. Chapman, Letcombe Laboratory.)

that *phototrophs* use light energy whereas *chemotrophs* use a chemical source of energy. *Lithotrophs* or *autotrophs* utilize carbon dioxide whereas *organotrophs* or *heterotrophs* utilize organic carbon compounds. *Chemoheterotrophs* are probably the most common group in soils.

Some important genera of bacteria which do (G+) and do not (G−) take up Gram stain present in the soil are *Arthrobacter* (G+ rods/cocci), *Azotobacter* (G− rods or yeast-like cells which form cysts and fix atmospheric dinitrogen when a fixed nitrogen source is unavailable), *Bacillus* (G+ rods with spores), *Clostridium* (G+ rods which are obligately anaerobic), *Chondromyces* (G− flexible rods with cysts), *Cytophaga* (G− flexible rods), *Micrococcus* (G+ cocci which are aerobic or micro-aerophilic), *Nitrosomonas* (G− rods which oxidize ammonia to nitrate and are lithotrophic), *Pseudomonas* (G− rods which often produce fluorescent pigments) and *Rhizobium* (G− rods which form nodules with legumes and fix atmospheric nitrogen). Some genera of actinomycetes in soil are *Micromonospora, Nocardia, Streptomyces* (which often produce antibiotics), *Streptosporangium* and *Thermactinomycetes* (which are thermophilic).

The soil algae include members of the divisions Cyanochloronta (blue-green algae), Chlorophycophyta (green algae), Chrysophycophyta (diatoms and yellow-green algae), Euglenophycophyta (euglenoids) and Rhodophycophyta (red algae) but most algae found in soil belong to the former two divisions. Classification of the blue-green algae [19] into taxonomic categories is not agreed upon and this is partly because it is difficult to obtain them in pure culture without inducing mutations. The systematics of eukaryotic algae is also complex [19]. Blue-green algae, such as *Scytonema* sp. (Fig. 3.4), have a significance for agriculture which is discussed further in section 10.8.

The groups of fungi in soil are Chytridiomycetes (*Allomyces, Rhizophydium*), Oomycetes (*Pythium, Rhizophydium*), Zygomycetes (*Absidia, Mortierella, Mucor, Rhizopus, Zygorhynchus*), Ascomycetes (*Chaetomium, Gymnoascus, Sordaria, Saccharomyces*), Basidiomycetes (*Boletus, Corticium, Marasimius, Omphalina, Tricholoma*), Deuteromycetes, the fungi imperfecti (*Arthrobotrys, Aspergillus, Cladosporium, Fusarium, Gliocladium, Pencillium, Trichoderma*), mycelia sterilia (*Rhizoctonia*) and Myxomycetes, the true slime moulds, which have some similarities to protozoa (*Physarum*).

The protozoa are animals but some contain chlorophyll. Vegetative cells of the class Mastigophorea possess whip-like locomotory organelles or flagella. This class contains photosynthetic algal flagellates, their colourless relatives and the so-called zooflagellates. In

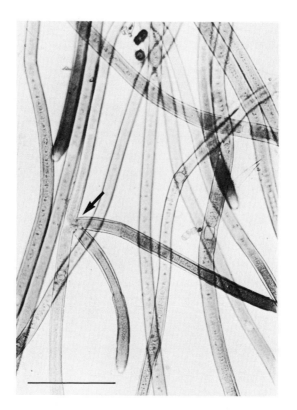

Fig. 3.4   Soil alga *Scytonema* sp. It forms algal crusts on the soil of the Sahel areas of Nigeria and is especially common after burning. Note the false branching pattern (arrowed) and the thick sheath. The latter, which has a polysaccharide nature, probably contributes to its action in stabilizing soils and increases moisture retention by the soil. As a dinitrogen fixer, it makes a contribution to soil nitrogen. Bar marker = 100 μm. (Photograph by W. D. P. Stewart, Dundee University.)

soil, the phytoflagellates are represented by the orders Crytomonadida, Chrysomonadida and Euglenida; the common zooflagellates present are in the Kinetoplastida and the Choanoflagellida. The classes Amoebida (amoebae) and Testacida (testate amoebae), Acrasida (myxamoebae or slime moulds) and Heliozoida all normally possess flexible external protrusions or pseudopodia, which are often used as locomotory organelles. The Gliophora (ciliates) can be distinguished by the bristle-like locomotory organelles or cilia on their body surface, the associated pellicular organization, their unique dimorphism and their method of conjugation. Inevitably these classes are represented by many genera [7].

Species such as *Acanthamoeba palestinensis* (Fig. 3.5) and *Vahlkampfia aberdonica* are common in soils although flagellates can be

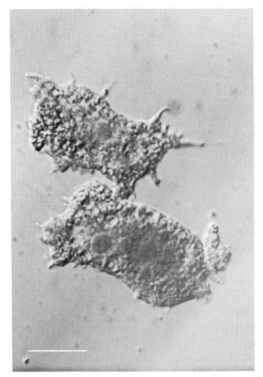

Fig. 3.5 Two trophozoites of the soil amoeba *Acanthamoeba palestinensis* (Reich) by differential interference contrast (Nomarski) microscopy. Bar marker = 10 μm. (Photograph by J. F. Darbyshire, Macaulay Institute for Soil Research.)

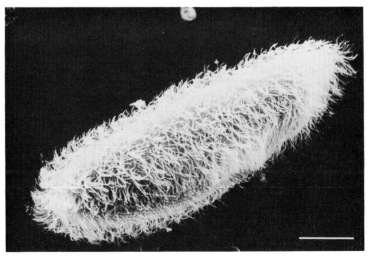

Fig. 3.6 Ciliate protozoan from soil. Bar marker = 20 μm. (Photograph by J. Bebb, Weed Research Organization.)

more common (Fig. 3.6). Probably the shape of *A. palestinensis* is what most people associate with amoebae but the single pseudopodium or limax type is the more usual. Although protozoa can be easily found in agricultural soils, their contribution to the biomass and overall biochemical activity is still uncertain. The protozoa, blue-green and higher algae, actinomycetes and viruses have received far less attention than bacteria and fungi from soil microbiologists. As such, we have far less idea of their relative contributions to the soil biomass and of their overall significance.

3.2   Methods of study

3.2.1   *Form and arrangement*

The light microscope and electron microscope, both scanning (SEM) and transmission (TEM), are useful instruments with which to study soil micro-organisms and each provides different types of information. With the electron microscope, greatest detail is provided by TEM but only very thin sections of roots or soil can be used; this means that it is more likely to encounter a bacterium than a fungus, because although the fungal biomass in soil may be greater, it is more localized. SEM is less powerful but provides a three-dimensional view. Both TEM and SEM suffer the disadvantage that chemical and physical treatments must be used in preparing a sample and this might introduce artefacts to the structure. The light microscope is particularly useful for counting micro-organisms. It is usual to disperse plant or soil samples with a suitable suspending fluid and then release the micro-organisms by shaking, bombarding with glass beads, blending, macerating, stomaching or treating with ultrasonics. Phenol aniline blue is a commonly used stain since it tends only to stain living cells. However, it is now possible to determine more positively the presence of living or dead cells with a differential fluorescent stain. This is a combination of two stains—a europium chelate and a fluorescent brightener—in which viable (or recently dead) cells fluoresce red and dead cells fluoresce green [14,15] (Figs 5.7 and 5.8). This technique, in common with other fluorescent stains such as acridine orange, is difficult to use. It needs a good microscope with fluorescence attachments and is difficult to interpret because colours fade rapidly; it is therefore of limited use for routine measurements. A recent staining method which detects the presence of metabolically active cells is the use of tetrazolium which is responsive to the dehydrogenase activity [18]. Nomarski differential interference microscopy also needs special microscope attachments but provides a three-dimensional effect with light microscopy; this technique is particularly useful for counting micro-organisms on root surfaces (Fig. 5.6). However, all techniques are difficult to use for *in situ* measurements because light may not penetrate and organisms can be masked.

To trace a specific micro-organism in soil, immunofluorescence is a powerful technique (Fig. 3.7). A specific micro-organism is injected into an animal such as a rabbit and an antibody is formed; this is then labelled with a fluorescent stain such as fluorescein isothiocyanate (FITC) which is bound to proteins [2]. This labelled antibody will then react with the antigen of the specific micro-organism in the soil and the micro-organism will take up the fluorescence. The technique has been used particularly to study rhizobia in soils [3].

To recover and count inoculated organisms from soil, antibiotic marking is sometimes used. The inoculum is grown on successively higher doses of antibiotic, such as streptomycin, in liquid or solid medium [16]. On inoculation and reisolation from the soil, the marked organisms will grow on agar containing the antibiotic, whereas other organisms will not. To be more precise, multiple antibiotic resistance is sometimes induced.

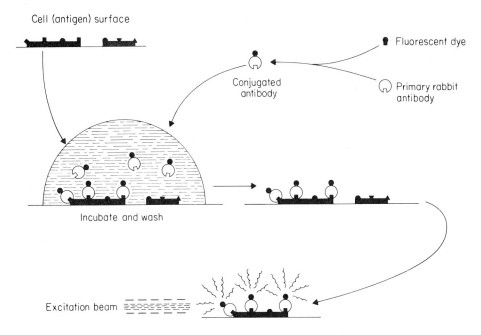

Fig. 3.7   Immunofluorescence technique (direct method).

3.2.2

Isolation

It must be recognized that in all isolation procedures, micro-organisms are taken from natural conditions and placed in artificial. This can affect their form, induce dormancy or even kill them. All techniques used are selective, e.g. the growth of fungi on agar is highly selective for sporulating forms. Bacterial isolations are equally selective and therefore plate counts account for only about 1–50 per cent of the total count.

3.2.3

Biomass measurements

The biomass (total weight of living microbial cells) can be assessed from microscopic measurements. The volume can be measured from the cell dimensions and the density by suspending cells in solutions of salts (e.g. CsCl) with different densities, so that cells with a similar density as that of the suspending solution cannot be sedimented or centrifuged. Caution should be exercised however in the interpretation of such data because the osmoticum may extract water from the cell giving an erroneously high result. Then:

Biomass = number of cells × density × volume.

A spherical bacterial cell of $1.0 \, \mu m$ diameter and density of $1.5 \, g \, cm^{-3}$ weighs *c.* $0.8 \times 10^{-12} \, g$. Fungal mycelium can be assumed to be cylindrical (volume $= \pi r^2 \, l$) with a diameter of *c.* $4 \, \mu m$. The density is often quoted as about $1.5 \, g \, cm^{-3}$ but the basis for this is unclear and it seems rather heavy; possibly errors may originate from density gradient centrifugation.

For a mixed microbial population in soil, these measurements are extremely difficult and tedious to make. However, Jenkinson and Powlson [12] made measurements for a wide range of pure cultures of micro-organisms in soil, then fumigated them with chloroform and measured the amount of cell carbon mineralized to carbon dioxide in the succeeding ten days. A mean figure (k) was obtained for the range of organisms and a value of 0.41 now seems to be most appropriate [1]. Then the biomass, B (mg C $100 \, g^{-1}$ dry soil), can be calculated from:

$$B = \frac{X - x}{k}$$

where X is the carbon dioxide produced by fumigated soil in ten days (mg $CO_2 - C$) and x is that produced by unfumigated soil. Others [17] have shown in their soils that:

$$B = \frac{0.673 \, X - 3.53}{k}$$

which avoids the need to measure carbon dioxide from unfumigated cores; however, they have also indicated that anomalous results can be obtained if soils are sieved prior to the determination. The amount of phosphorus held in soil micro-organisms can be calculated from the difference between the amount of inorganic phosphorus extracted by $0.5 M - NaHCO_3$ (pH 8·5) from fresh soil fumigated with $CHCl_3$ and the amount extracted from unfumigated soil [4]. This provides an alternative to the measurement of biomass carbon.

Another method is to measure the ATP content of soil (Fig. 3.8); ATP is required for the biosynthetic and catabolic reactions of cells. The major variant of this technique, as used by different investigators, is the extractant used; with Na_2HPO_4 and paraquat dichloride, Jenkinson and Oades [13] related their estimates to those of the fumigation method by:

Biomass C in soil $= K$ (ATP content of soil)

where K is 120, although recent evidence shows 160 to be a better value. The correlation may break down in soils that have recently received large additions of substrate, particularly as phosphate is in short supply. An improved method for ATP measurement is where a purified rather than a crude luciferase preparation is used [25].

The heat output of soils can be assessed by microcalorimetry. Heat output correlates well with respiration but to a lesser extent with biomass [23]. Thus, this is an additional technique for assessing the overall biological activity of a soil but normally it would be simpler to merely measure respiration. All methods of biomass estimation, such as the measurement of muramic acid, chitin and nucleic acid,

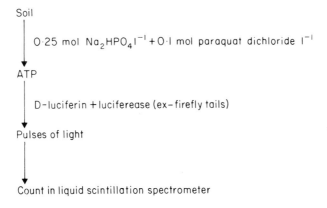

Fig. 3.8 ATP extraction from cells. (After Jenkinson & Oades [13].)

were the subject of extensive debate at the Second International Symposium on Microbial Ecology and it seems unwise to regard any of the methods as absolute. A recent comprehensive review is available [11].

Soil enzymes

The total biochemical activity of soil is comprised of a series of reactions catalysed by enzymes. These may be intracellular within living or dead organisms or extracellular, sometimes distant from their point of origin (Table 3.2). Extracellular enzymes can be free or bound to soil colloids. The text edited by Burns [6] is a useful source of reference.

More than 50 enzymes have been shown to be active in soil (Table 3.3), including oxidoreductases, hydrolases and transferases. In some investigations a clear correlation has been shown between microbial numbers and enzyme activity, but sometimes activity appears to be independent of microbial proliferation. However, microbial numbers seldom correlate with microbial biomass, which is a better index of microbial activity, and there have been few attempts to correlate biomass with enzyme activity. It is quite clear that a proportion of enzyme activity is independent of living cells because treatment of soil with toluene, sodium azide or gamma rays usually destroys micro-organisms and extracellular unbound enzymes without preventing enzyme activity.

The study of soil enzymes is clearly more difficult than studies of the pure enzymes that biochemists are familiar with and where the disappearance of substrate and appearance of products is monitored. Unless the duration of assay is brief (less than two hours) it is desirable to eliminate the complications of microbial growth as a contributor to the activity; authorities differ on the most satisfactory way to achieve this because there may be simultaneous cell lysis. Temperature, pH, moisture and agitation must also be considered in standardizing assays, in the same way as for 'conventional' enzymology. However, the sorptive influence of soil colloids (Chapter 2) is probably the major complicating factor.

Microbial decomposition processes in soil (Chapter 4) initially involve large polymers such as lignocelluloses and the first steps are brought about by exoenzymes, even though later stages may be intracellular. The cellulase complex of enzymes (Fig. 3.9) is still relatively poorly understood, although rapid advances are being made now that the potential of this decomposition is recognized for biotechnology (Chapter 10). It used to be considered that a C_1-cellulase (also known as exo-glucanase or exo(syl)hydrolase, E.C.

Table 3.2 Conceptual scheme of the components of soil enzyme activities. (The major components found experimentally in soils are capitalized.) (From Skujins [22].)

	Enzymatic activity in soil							
	Abointic enzymes							Endocellular enzymes of proliferating micro-organisms, plant roots and soil fauna
	Accumulated enzymes					Continuously released extracellular enzymes		
	Bound to microbial cellular components		Not associated with cellular components			From micro-organisms	From plant	
	In intact dead cells	In cellular fragments	Originating from micro-organisms and soil fauna		Originating from plant roots	organisms	roots	
Origin			Endocellular enzymes from disrupted cells	Extracellular enzymes				
Location in soil	IN NON-PROLIFERATING CELLS		BOUND TO SOIL COMPONENTS			In liquid phase		IN ORGANISMS

Table 3.3   Enzymes in soil. (After Thornton & McLaren [26].)

Enzyme	Reaction catalysed
Oxidoreductases	
Catalase	$2H_2O_2 \rightarrow 2H_2O + O_2$
Catechol oxidase (tyrosinase)	o-diphenol $+ \frac{1}{2}O_2 \rightarrow o$-quinone $+ H_2O$
Dehydrogenase	$XH_2 + A \rightarrow X + AH_2$
Diphenol oxidase	p-diphenol $+ \frac{1}{2}O_2 \rightarrow p$-quinone $+ H_2O$
Glucose oxidase	Glucose $+ O_2 \rightarrow$ gluconic acid $+ H_2O_2$
Peroxidase and polyphenol oxidase	$A + H_2O_2 \rightarrow$ oxidized $A + H_2O$
Urate oxidase (uricase)	Uric acid $+ O_2 \rightarrow$ allantoin CO_2
Transferases	
Transaminase	$R_1R_2\text{-CH-N}^+H_3 + R_3R_4CO \rightarrow$
	$R_3R_4\text{-CH-N}^+H_3 + R_1R_2CO$
Transglycosylase and levansucrase	$nC_{12}H_{22}O_{11} + ROH \rightarrow$
	$H(C_6H_{10}O_5)nOR + nC_6H_{12}O_6$
Hydrolases	
Acetylesterase	Acetic ester $+ H_2O \rightarrow$ alcohol $+$ acetic acid
α- and β-amylase	Hydrolysis of 1,4-glucosidic bonds
Asparaginase	Asparagine $+ H_2O \rightarrow$ aspartate $+ NH_3$
Cellulase	Hydrolysis of β-1,4-glucan bonds
Deamidase	Carboxylic acid amide $+ H_2O \rightarrow$ carboxylic acid $+ NH_3$
β-fructofuranosidase (invertase, sucrase, saccharase)	β-fructofuranoside $+ H_2O \rightarrow ROH +$ fructose
α- and β-galactosidase	Galactoside $+ H_2O \rightarrow ROH +$ galactose
α- and β-glucosidase	Glucoside $+ H_2O \rightarrow ROH +$ glucose
Inulase	Hydrolysis of β-1,2-fructan bonds
Lichenase	Hydrolysis of β-1,3-cellotriose bonds
Lipase	Triglyceride $+ 3H_2O \rightarrow$ glycerol $+ 3$ fatty acids
Metaphosphatase	Metaphosphate $\rightarrow$ orthophosphate
Nucleotidase	Dephosphorylation of nucleotides
Phosphatase	Phosphate ester $+ H_2O \rightarrow ROH +$ phosphate
Phytase	Inositol hexaphosphate $+ 6H_2^3O \rightarrow$ inositol $+ 6$ phosphate
Protease	Proteins $\rightarrow$ peptides and amino acids
Pyrophosphatase	Pyrophosphate $+ H_2O \rightarrow 2$ orthophosphate
Urease	Urea $\rightarrow 2NH_3 + CO_2$

3.2.1.91) brought about the preliminary attack, usually involving solubilization. Subsequent hydrolysis then depended on a C_x-cellulase (also known as endoglucanase or 1,4-β-glucano-hydrolase, E.C. 3.2.1.4). However, it is now known that C_1-cellulase has a low affinity for cellulose and probably only degrades highly crystalline celluloses. C_x-cellulase attacks insoluble, crystalline celluloses as well as soluble forms of cellulase such as carboxymethylcellulose. It has been

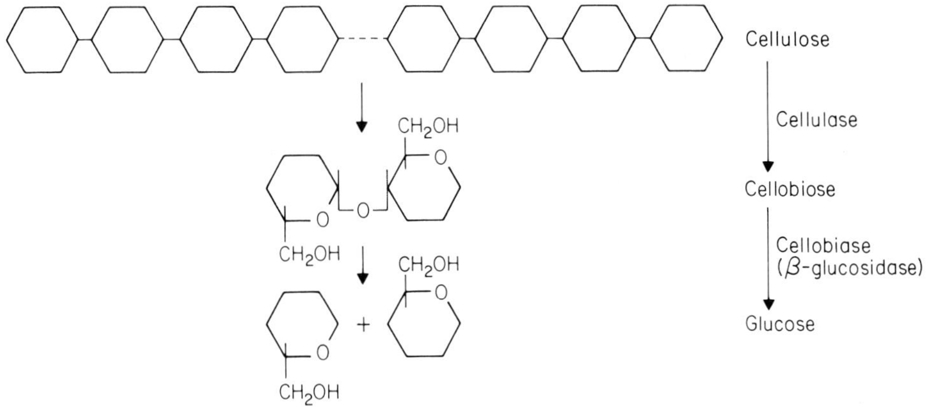

Fig. 3.9   Breakdown of cellulose.

suggested [29] that the enzymes operate synergistically and, indeed, that the C_x-cellulase produces the fragments required for C_1-cellulase activity. Thus, their sequence of action appears to be the reverse of the originally proposed mechanism. Both cellulases produce cellobiose which is degraded to glucose monomers by cellobiase (also known as 1,4-β-glucosidase or β-D-glucosidase, E.C. 3.2.1.21).

The nitrogen cycle (Chapter 4) has major agricultural, economic and ecological relevance. The enzyme responsible for nitrogen fixation—nitrogenase—has now been studied extensively but urease has also received considerable attention. The substrate urea occurs naturally from animal excreta and as a product of nucleic acid mineralization but it is also an increasingly important nitrogen fertilizer:

$$NH_4CONH_2 \xrightarrow[H_2O]{} NH_2COOH + NH_3 \xrightarrow[H_2O]{} CO_2 + 2NH_3.$$

The product—ammonia—is oxidized by chemolithotrophic bacteria to nitrate; both ammonium salts and nitrate can be absorbed by plant roots.

Urease preparations from soils [20] are quite different from pure enzyme systems (e.g. Jack bean urease), being resistant to proteolytic activity, able to withstand high temperatures and little affected by long-term storage. The association of the enzyme with soil organic matter changes both the maximum velocity of the enzyme–substrate interaction (V_{max}) and the affinity of the enzyme for its substrate (K_m—the Michaelis constant).

3.4   **Conclusion**

A diverse range of micro-organisms, with their associated enzymes, and free enzymes exist in soils. Modern methodology now provides us with the opportunity to make qualitative and quantitative assessments of the soil biomass and its activity. Application of these modern techniques, in association with classical methodology, will enable the significance of micro-organisms to soils and plants to be assessed.

References

1 Anderson J. P. E. & Domsch K. H. (1978) Mineralization of bacteria and fungi in chloroform-fumigated soils. *Soil Biology and Biochemistry*, **10**, 207–13.

2 Babiuk L. A. & Paul E. A (1970) The use of fluorescein isothiocyanate in the determination of the bacterial biomass of grassland soil. *Canadian Journal of Microbiology*, **16**, 57–62.

3 Bohlool B. B. & Schmidt E. L. (1980) The immunofluorescence approach in microbial ecology. *Advances in Microbial Ecology*, **4**, 203–41.

4 Brookes P. C., Powlson D. S. & Jenkinson D. S. (1982) Measurement of microbial biomass phosphorus in soil. *Soil Biology and Biochemistry*, **14**, 319–29.

5 Buchanan R. E. & Gibbons N. E. (eds) (1974) *Bergey's Manual of Determinative Bacteriology*, 8th edn. Williams & Wilkins, Baltimore.

6 Burns R. G. (ed.) (1978) *Soil Enzymes*. Academic Press, London.

7 Darbyshire J. F. (1975) Soil protozoa-animalcules of the subterranean microenvironment. In *Soil Microbiology. A Critical Review*, ed. Walker N., pp. 147–63. Butterworth, London.

8 Domsch K. H. & Gams W. (1972) *Fungi in Agricultural Soils*. Longman, London.

9 Domsch K. H., Gams W. & Anderson T-H. (1982) *Compendium of Soil Fungi*, vols 1 and 2. Academic Press, London.

10 Holt J. G. (ed.) (1977) *The Shorter Bergey's Manual of Determinative Bacteriology*, 8th edn. Williams & Wilkins, Baltimore.

11 Jenkinson D. S. & Ladd J. N. (1981) Microbial biomass in soil: measurement and turnover. In *Soil Biochemistry*, vol. v, eds Paul E. A. & Ladd J. N., pp. 415–71. Marcel Dekker, New York.

12 Jenkinson D. S. & Powlson D. S. (1976) The effects of biocidal treatments on metabolism in soil. V. A method for measuring soil biomass. *Soil Biology and Biochemistry*, **8**, 209–13.

13 Jenkinson D. S. & Oades J. M. (1979) A method for measuring adenosine triphosphate in soil. *Soil Biology and Biochemistry*, **11**, 193–9.

14 Johnen B. G. (1978) Rhizosphere micro-organisms and roots stained with europium chelate and fluorescent brightener. *Soil Biology and Biochemistry*, **10**, 495–502.

15 Johnen B. G. & Drew E. A. (1978) Morphology of micro-organisms stained with europium chelate and fluorescent brightener. *Soil Biology and Biochemistry*, **10**, 487–94.

16 Ko W. H. & Chow F. K. (1977) Characteristics of bacteriostasis in natural soils. *Journal of General Microbiology*, **192**, 295–8.

17 Lynch J. M. & Panting L. M. (1982) Measurement of the microbial biomass in intact cores of soil. *Microbial Ecology*, **7**, 229–34.

18 MacDonald R.M. (1980) Cytochemical demonstration of catabolism in soil micro-organisms. *Soil Biology and Biochemistry*, **12,** 419–23.

19 Metting B. (1981) The systematics and ecology of soil algae. *Botanical Review*, **47,** 195–312.

20 Pettit N.M., Smith A.R.J., Freedman R.B. & Burns R.G. (1976) Soil urease: activity, stability and kinetic properties. *Soil Biology and Biochemistry*, **8,** 479–84.

21 Sieburth J.McN. (1977) How can we divide the microbes? In *CRC Handbook of Microbiology, Vol. 1 Bacteria*, 2nd edn, eds Laskin A.T. & Lechavallier H.A., pp. 3–7. CRC press, Cleveland, Ohio.

22 Skujins J.J. (1967) Enzymes in soil. In *Soil Biochemistry*, vol. 3, eds McLaren A.D. & Peterson G.H., pp. 317–414. Marcel Dekker, New York.

23 Sparling G.P. (1981) Microcolorimetry and other methods to assess biomass and activity in soil. *Soil Biology and Biochemistry*, **13,** 93–8.

24 Stainier R.Y. & Van Niel C.B. (1962) The concept of a bacterium. *Archiv fur Mikrobiologie*, **42,** 17–35.

25 Tate K.R. & Jenkinson D.S. (1982) Adenosine triphosphate measurement in soil: an improved method. *Soil Biology and Biochemistry*, **14,** 331–5.

26 Thornton J. & McLaren A.D. (1975) Enzymatic characterization of soil evidence. *Journal of Forensic Science*, **20,** 674–92.

27 Whittaker R.H. (1969) New concepts of kingdoms of organisms. *Science*, **163,** 150–60.

28 Winogradsky S. (1924) Sur la microflore autochtone de la terre arable. *Comptes rendus ebdomadaire des séances de l'Academie des Sciences* (*Paris*) *D*, **178,** 1236–9.

29 Wood T.M. & McCrae S.I. (1978) The mechanism of cellulose action with particular reference to the C_1 component. In *Bioconversion of Cellulosic Substances into Energy, Chemicals and Microbial Protein*, ed. Ghose T.K., pp. 111–41. Indian Institute of Technology, Delhi.

MICROBIAL SAPROPHYTES: DECOMPOSITION PROCESSES AND NUTRIENT CYCLES

Recently further important information has been discovered about decomposition processes. It has been proved that these processes, so similar to fermentations, are governed by simple organisms, and indeed the lowest forms of the plant world, the fungi and bacteria. It will be necessary, in the future, to determine not merely decomposition products but also the causes of these transformations, the micro-organisms themselves ... if we are to obtain an adequate idea of their ill-effects on health.

R. Koch *Deutsches Artzblatt*, cxxxvii, 244–50, 1882.

Saprophytes are important members of the decomposer communities which bring about the cycling of nutrients. However, it is important to recognize that in the nutrient cycles they interact with all other nutritional groupings, such as the symbiotes (Chapter 6). Decomposition in terrestrial ecosystems is the subject of a detailed monograph [23] which adopts the ecological energetics approach. Another book is available on the role of bacteria in mineral cycling [7]; this adopts the biochemical energetics approach.

4.1   Sources of substrates

The decomposition of organic materials released by living plant, animal or microbial cells and the breakdown of dead organisms is brought about by microbial saprophytes. Table 4.1 illustrates the amount of these materials available in the surface 5 cm of an arable soil. The non-cellular native soil organic matter or humus (Fig. 4.1) is not a very satisfactory substrate because it has commonly been dated with ^{14}C at a half-life in excess of 1000 years [3]. The recalcitrance is probably due, at least in part, to spatial inaccessibility. Humic and fulvic acids are extremely heterogeneous cross-linked

Table 4.1   Approximate annual input of substrates (primary productivity) to the surface 5 cm of an arable soil (with an approximate microbial biomass of 400 kg C ha^{-1}). (From Lynch & Panting [14].)

Source	Amount (kg C ha^{-1} year^{-1})
Root decomposition	400
Root exudation	240
Straw residues	2800
Autotrophic microbes	100
Total	3540

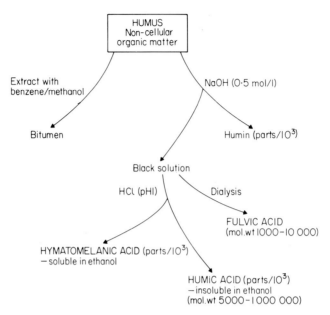

Fig. 4.1 Fractionation of humus.

polymers with central cores. The extracted, and so probably modified, polymers can be degraded by soil micro-organisms [11,15]. The significance of this in nature is unclear and indeed it is possible that the heterogeneous nature of the polymers would make them chemically recalcitrant because of the difficulty in forming enzyme-substrate complexes.

Straw provides the major available substrate in the surface layer of arable soil. The contribution of straw to the substrate pool below the plough layer is very small. In deeper horizons, root exudation and decomposition of dead roots is relatively more important. The structural organization of straw is shown in Fig. 4.2. There is layering of lignocellulose around the nodes, which causes a greater rate of decomposition compared with the remainder of the straw [9]. Although the structure of straw has been studied little in relation to its degradation by micro-organisms, this must be considered in more detail if plant residues are to be harnessed in useful biological processes (Chapter 10).

Any plant residue or tissue can be fractionated into the following components: cellulose (a polymer of glucose), hemicellulose (a polymer of glucose and other sugars), lignin (a polymer of phenols), protein, water-soluble materials (such as sugars) and ether-soluble materials (such as lipids). They can be assayed by proximate analysis— the sequential removal of each fraction from the tissue. Straw

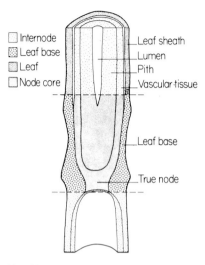

Fig. 4.2 Structure of straw. (From Harper & Lynch [9].)

principally contains cellulosic materials and lignin (lignocellulose) but their precise chemical and physical association in the straw is uncertain at present. Fresh plant tissues, such as the rhizomes of couch grass, have a greater proportion of non-lignocellulosic materials, such as free sugars (Table 4.2).

Table 4.2 Proximate analysis of wheat straw and couch (quack)-grass rhizomes.

Tissue	Cellulose	Hemi-cellulose	Lignin	Protein-soluble materials	Water-soluble materials	Ether
			(% dry weight)			
Wheat straw	43	36	14	1	4	3
Couch-grass rhizomes	18	23	12	3	39	1

4.2   **Kinetics of decay**

The decomposition of plant material in soil can be studied by measuring weight loss directly (Fig. 4.3) or by monitoring the $^{14}CO_2$ produced from ^{14}C-labelled material (Fig. 4.4). The decay curves produced are usually bi- or polyphasic. Many environmental and physical factors govern the kinetics of decay. The sites in California in 1977 (Fig. 4.3) were close together and yet the decay rates were quite different.

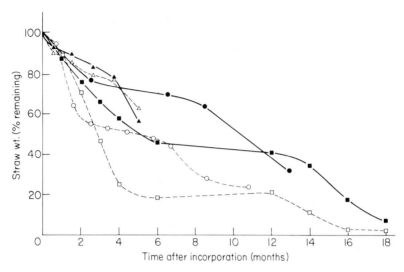

Fig. 4.3   Straw decomposition as measured by weight loss. Solid circle = wheat straw from Kimberly, Idaho [22]; solid square = wheat straw from Bozeman, Montana [2]; open square = wheat-straw from Huntley, Montana [2]; solid triangle = rice straw from Davis, California [18]; open triangle = rice straw from Knight's Landing, California [18]; open circle = oat straw from Northfield, Oxfordshire [10].

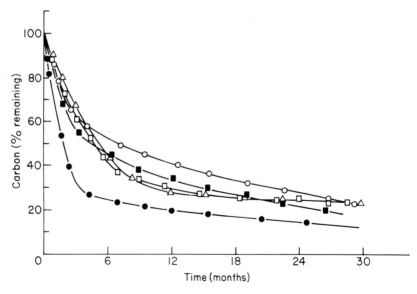

Fig. 4.4   Decomposition of ^{14}C-labelled plant residues in soil. Open circle = wheat straw from Chernozemic, Canada [20]; solid square = wheat straw from Luvisolic, Canada [20]; open triangle = wheat straw from West Germany [19]; solid circle = rye-grass from Nigeria [12]; open square = rye-grass from England [12].

Even though the decay is seldom monophasic, results are usually analysed by first-order rate kinetics:

$$\frac{-dA}{dt} = kA$$

where A is the concentration of the added organic matter, t is time and k is the decomposition-rate constant (time^{-1}). On integration:

$$A = A_0 e^{-kt} \text{ or } Ln\frac{A}{A_0} = kt$$

where A_0 is the concentration of A at zero time. Values of k can be determined by plotting $Ln\, A/A_0$ against t. The technique using ^{14}C material can be more difficult to interpret in this respect because some of the ^{14}C is incorporated into microbial cells and metabolites; therefore not all the degraded ^{14}C is measured as $^{14}CO_2$. Weight loss *per se* is usually determined from material contained within nylon mesh bags. Preferably the bags should contain some soil to provide an even inoculum and different sizes of mesh should be employed to separate the contributions of animals and micro-organisms in the decomposition.

Lignin is spatially inaccessible and also probably chemically recalcitrant to micro-organisms during the early stages of decomposition of plant residues and therefore cellulose and hemicellulose disappear first (Table 4.3). However, some cellulosic materials are associated with lignin and therefore they also are somewhat recalcitrant. Lignin decays very little during the first year after incorporation into soil and therefore analyses of the lignin percentage in the decaying material is a good index of the state of decomposition [10]. A better understanding of lignin decomposition will be critical to the full utilization of lignocellulose wastes (Chapter 10). Indeed, at present, the partial structure of the polymer has only been evaluated for spruce lignin and there has been no attempt to analyse straw lignin. A few laboratories are now actively investigating the microbiology and biochemistry of lignin degradation; some are even investigating the potential to produce mutants which carry out the process more efficiently and *Streptomyces* spp. [1] have been the organisms of choice for this.

Table 4.3 Decomposition rate constants (k) for the major components of buried wheat straw measured September 1977–August 1978. (After Harper & Lynch [10].)

Components	k (d^{-1})
Hemicellulose	0·0073
Cellulose	0·0074
Lignin	0·0005

Microbial growth rates

Whereas analysis of decomposition concerns principally the substrate disappearance, it is appropriate to consider microbial growth dynamics when interpreting decomposition.

A classic microbial growth curve is shown in Fig. 4.5. Growth

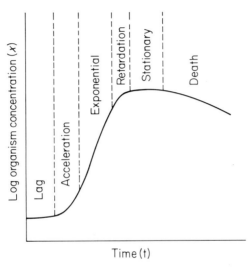

Fig. 4.5 Theoretical microbial growth curve.

takes place during the exponential phase, while carbon substrates are available and there is no other nutrient limitations. The growth rate is the maximum that can be achieved by the organism. During exponential growth, if there are x individuals initially, the following applies:

Number of generations	0	1	2	3	4	5	n
Number of individuals	x	$2x$	$4x$	$8x$	$16x$	$32x$	ax
	2^0x	2^1x	2^2x	2^3x	2^4x	2^5x	2^nx.

The first law of growth is then expressed as:

$$dx/dt = \mu x$$

where μ is a constant termed the specific growth rate.

If there are x_t organisms at time t, integrating and using the natural logarithms:

$$Ln\ x_t = Ln\ x_0 + \mu t \text{ and } Ln\ ^{x}t/x_0 = \mu t$$

when the population doubles in size,

$$x_t/x_0 = 2 \text{ and } t = t_d$$

t_d is termed the doubling time.

Then:

$Ln2 = \mu t_d.$

If the initial inoculum concentration is x, the growth yield, Y, can be found by knowing the amount of substrate consumed, ΔS. The second law of growth is:

$Y = (x - x_0)/\Delta S$ or $Y = -dx/dS.$

The metabolic quotient, q, can be calculated from:

$dS/dt = -qx$ and $-dS = qxdt.$

However, as made clear by the first and second laws:

$-dS = (\mu xdt)/Y$ and therefore $q = \mu/Y.$

The considerations so far refer to growth in a closed system, i.e. there is no fresh input of substrates or dilution of the culture. In nature, the system is more likely to be open and somewhat analogous to a continuous culture (chemostat) as follows:

One volume is displaced in a time, θ (the retention time):

$\theta = V/F.$

If the dilution rate is D (h^{-1}),

$D = F/V.$

Various events can take place when the substrate flow is started (Fig. 4.6).

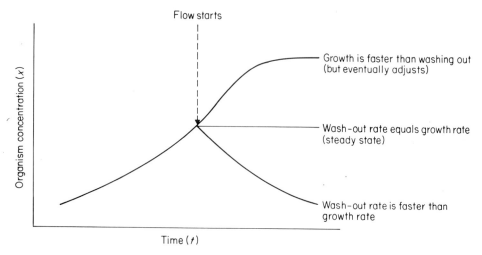

Fig. 4.6   Effect of substrate flow on a growing microbial population.

Change in organism concentration = growth − outflow.

For a unit volume of culture in the interval dt:

$dx = \mu x dt - Dx\ dt$ and $dx/dt = \mu x - Dx$ and $dx/dt = (\mu - D)\ x$.

In the steady state:

$dx/dt = O$ and $\mu = D$.

This does not allow for the energy used in the maintenance of the cell, which is important when μ is small.

Rate of substrate utilization	=	rate of substrate used for biomass production	+	rate of substrate used for maintenance
dS/dt	=	$\mu x / Y_G$	+	mx

where Y_G is the true growth yield when no substrate is used in maintenance energy. S can generally be calculated, x can be determined and m is known for many organisms. For most of the analyses which have been made so far, mx has been greater than ds/dt. It is likely that the m values are too large and indeed m decreases with growth rate [4] (Fig. 4.7); it is difficult therefore to predict m when growth is very slow in soil. Furthermore, some organisms may not be dependent on exogenous energy as they may have accumulated

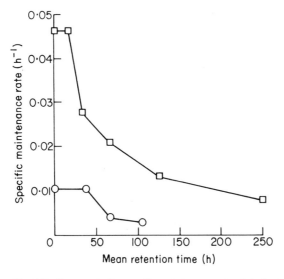

Fig. 4.7 Decrease in the specific maintenance rate of *Arthrobacter globiformis* (open circle) and *Klebsiella pneumoniae* (open square) in the chemostat with long retention times (low growth rates). (From Chapman & Gray [4].)

their own energy reserves and other organisms form resting structures, such as cysts, which probably have an extremely small maintenance requirement. The growth theory as described does not allow for the release of substrates during cell death and then utilization by survivors. Whereas this may be of minimal significance to rapidly growing populations on laboratory media, it is likely to be significant in the nutrient-limiting conditions of soil.

4.4   **Computer models**

Despite the difficulty of applying growth theory to organisms in soil, many computer models have attempted to predict the fate of carbon and nitrogen in the soil. A simple word model [16] is shown in Fig. 4.8. In the model, carbon dioxide evolution results from the decomposition of a single compound and the recycling of metabolites (section 4.2) affects the apparent decomposition rate as measured. The rate constants, k, were set at $0.08\,d^{-1}$ for the compound, $0.05\,d^{-1}$ for the decay of biomass and $0.04\,d^{-1}$ for the metabolite decay; all were assumed to follow first-order kinetics. Maintenance energy was set at 40 per cent of the carbon used by the biomass, the remainder being released as products. Efficiency of carbon use was assumed as

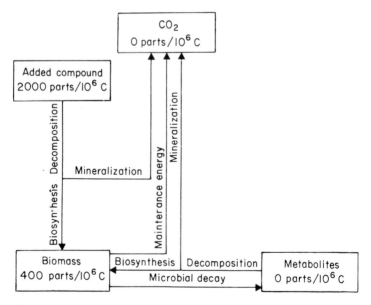

Fig. 4.8 Scheme of a model for the decomposition of a single compound. The figures in the boxes refer to the initial concentrations present. (From Paul & van Veen [16].)

60 per cent. The model was long-term, allowing for carbon turnover over hundreds of years, and placed the requirements for maintenance into a death coefficient. In this situation, much of the maintenance of the populations should come from cannibalism. Therefore, an increase in the rate of death would provide an increased supply of energy to the survivors. The output predicted from the computer is shown in Fig. 4.9. The true decomposition allows for metabolite production. Word models which can be translated into computer models have been constructed to allow for the different organic-matter fractions of soil, recalcitrance and spatial inaccessibility (Fig. 4.10). Even then, however, not all factors controlling decomposition have been considered and it must be borne in mind that nutrient cycles of the different elements (section 4.6) are interdependent, so that it is often difficult to determine the limiting nutrient. An additional limitation on such models is that we have too little information on the factors which must be assumed. Furthermore, the model can only gain credibility when it has been fully tested experimentally. Such experiments often point to the need for more informed additional experimentation.

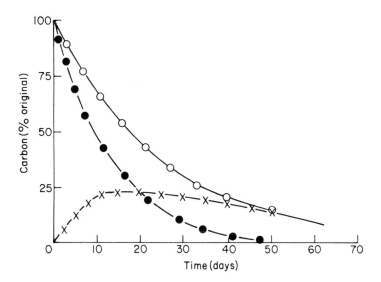

Fig. 4.9 Calculated decomposition of carbon added to the soil as a single compound that is cellulose. Open circle = total carbon remaining as calculated from carbon dioxide evolution; solid circle = true decomposition of carbon; cross = microbial production. (From Paul & van Veen [16].)

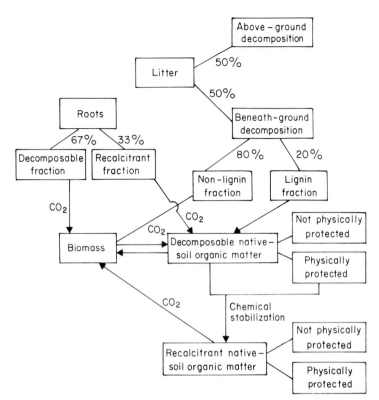

Fig. 4.10 Scheme of a more complex, long-term model of carbon decomposition in soil. (From Paul & Voroney [17].)

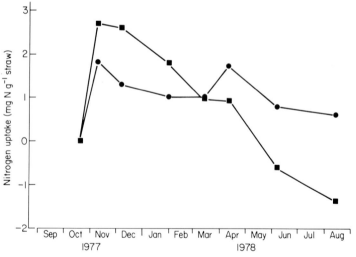

Fig. 4.11 Net immobilization/release of nitrogen during straw decomposition. The straw was left in the drill slit (solid circle) or buried (solid square) on 21st October. (Data of S. H. T. Harper and J. M. Lynch.)

Carbon/nitrogen ratio

In soils, decomposition of organic matter is commonly limited by nitrogen availability, e.g. straw has a carbon/nitrogen ratio of about

80:1. Therefore, 100 g straw contains about 40 g carbon and 0·5 g nitrogen. If we assume that 66 per cent of the carbon and nitrogen is available in six months, that the mean carbon/nitrogen ratio of micro-organisms is 5:1, that 35 per cent of carbon used (the remainder going to carbon dioxide) and 100 per cent nitrogen used enters the biomass, then an extra 1·5 g nitrogen 100 g^{-1} straw needs to be added to satisfy the carbon/nitrogen ratio of the biomass. The figures are, of course, a gross over-simplification as they ignore many factors such as the potential for recycling the nitrogen from dying cells, e.g. Fig. 4.11 shows that whereas nitrogen is temporarily immobilized by straw in the autumn, preventing fertilizer leaching during wet winters, it can be released in the spring when it may be of more value to the crop. Such calculations serve to emphasize that both carbon and nitrogen are important in decomposition and other elements, such as phosphorus and even the trace elements, can be limiting in some situations.

4.6   Nutrient (mineral) cycles

A simple approach to a part of the carbon cycle was illustrated in Fig. 4.8. Such cycles can be constructed for all biological elements, both in specific situations and in more global terms. Moreover, the cycles can be constructed at the biochemical and the ecological/ agricultural level. Two illustrative schemes for the nitrogen cycle are shown in Figs 4.12 and 4.13; such cycles are continually being modified as more information becomes available.

Reliable budgets of nitrogen cycling have been few but using ^{15}N-labelled fertilizer in lysimeters (with intact cores of soil 80 cm in diameter $\times$ 135 cm deep) it has been demonstrated [6] that crops take out a similar amount of nitrogen from the soil to that supplied in fertilizers (Table 4.4). This is because the crop extracts the nitrogen contained in the soil organic matter. Similarly, leaching losses are almost as great from the native soil organic matter as from the fertilizer. Losses of nitrogen as nitrous oxide produced during denitrification is rarely of agricultural significance but the losses of dinitrogen by the same process are more difficult to ascertain. The fraction of nitrogen in soil remains fairly constant because as nitrogen is removed, organic matter is simultaneously oxidized. Thus, the more important consequence of modern intensive cropping is that there is an increased tendency towards erosion as a result of the extra carbon lost. Although this is of small significance in countries such as Britain where soils generally have stable structures, in many countries this decline in soil as a natural resource could become of major consequence towards the end of the century, if not before.

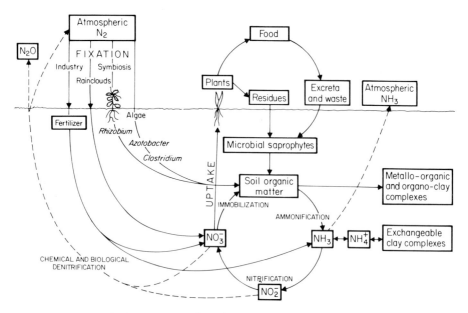

Fig. 4.12 Ecology of the nitrogen cycle.

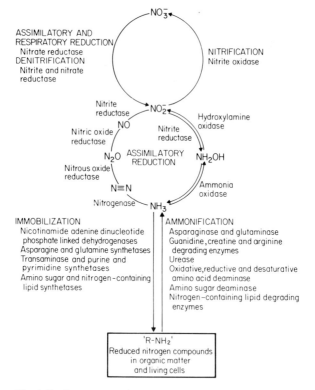

Fig. 4.13 Enzymology of the nitrogen cycle.

Table 4.4 (a) Nitrogen uptake into harvested parts of a grass crop for a year in which ^{15}N-labelled fertilizer was applied to lysimeters. (b) Loss of nitrogen by leaching in the winter after the fertilizer addition. (Data of R. J. Dowdell, Letcombe Laboratory.)

	Clay	Silt loam
	(kg N ha^{-1})	
Labelled nitrogen applied as Ca (NO$_3$)$_2$	400	400
(a) Nitrogen uptake by crop		
Labelled nitrogen	183	184
Unlabelled nitrogen	141	145
Total	324	329
(b) Loss of nitrogen by leaching		
Labelled nitrogen	25	22
Unlabelled nitrogen	19	16
Total	44	38

Besides carbon and nitrogen cycles, the phosphorus cycle (Fig. 4.14) is particularly relevant to plant growth. Mycorrhizae (section 6.3) can act as vectors for the uptake of phosphorus from the soil solution to plants, whereas rhizosphere bacteria (section 5.4) can promote or inhibit this process. Microbial transformations of organic phosphorus often produce the majority of solution phosphorus available to plants and this could be a beneficial effect of composting and organic farming (section 10.7). Phosphorus exists in the soil almost entirely as forms of phosphate and although it is present in soils at around 400–1200 mg kg^{-1}, little (probably less than 5 per cent) of this is available for plant growth, occurring mainly as insoluble phosphates and organic complexes. Although some phosphate enters the soil solution from the mineralization of rocks, the decomposition of organic matter is a more important source. The decay of nucleic acids and phospholipids is rapid but the more complex compounds, such as phytic acid and polyphosphates, are mineralized more slowly. The decomposer fungi contain about 0·5–1 per cent w/w; higher plants contain between 0·5–5 per cent w/w.

Although the nitrogen and phosphorus cycles have attracted major attention from microbiologists and soil scientists, cycles of the other elements can be of great consequence to microbial and plant growth. Many of the trace elements, such as iron, manganese, molybdenum and cobalt, are important for enzyme function and a deficiency of these in soil could limit useful activities such as nitrogen fixation. Nutrient deficiencies in soils can also change the populations of micro-organisms present, e.g. *Thiobacillus ferrooxidans* is responsible for the conversion of ferrous to ferric iron and also oxidizes sulphide

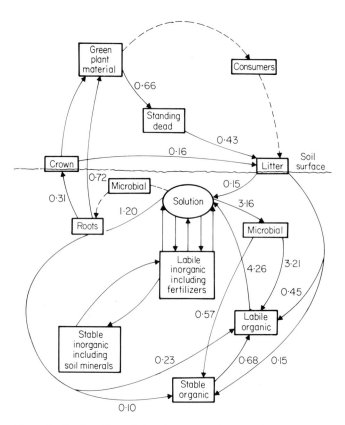

Fig. 4.14 Predicted flow of phosphorus (g m^{-2} year^{-1}) between components of a native grassland ecosystem. (After Cole *et al.* [5].)

to sulphate. However, the pH optimum for this process is very low (1·7–3·5) and therefore it has more geochemical than agricultural significance. Sulphur is not usually a limiting nutrient in soils; sulphur cycling is of primary concern when the soil redox potential is decreased and sulphate reduction occurs (section 7.3). The other major plant nutrient added as fertilizer—potassium—has received scant attention from the microbiologist, although the role of bacteria in its uptake by plants (section 5.4) has been studied.

Table 4.5 summarizes biomass and nutrient-flow data for an arable soil from Rothamsted which has not been manured or received fertilizer and has carried wheat every year since 1843 [13]. Some figures are measurements, others are estimates. The nitrogen and phosphorus flux through the biomass is roughly equivalent to the annual offtake in grain plus straw and suggests that the biomass pool is capable of supplying the major part of the nitrogen taken up by an

Table 4.5 The microbial biomass and related measurements in an unmanured soil under continuous wheat. (After Jenkinson & Ladd [13].)

Measurement	Value
Weight of soil	$2200 \, t \, ha^{-1}$
Organic matter in soil	$26 \, t \, C \, ha^{-1}$
Nitrogen in soil	$2 \cdot 7 \, t \, N \, ha^{-1}$
Annual input of organic matter	$1 \cdot 2 \, t \, C \, ha^{-1} \, year^{-1}$
Gross turnover time of soil organic carbon	22 years
Radiocarbon age of soil organic carbon (1944 sample)	1310 years
Number of 'spherical' organisms	$1100 \times 10^6 \, g^{-1}$
Volume of 'spherical' organisms	$0 \cdot 71 \, mm^3 \, g^{-1}$
Number of hyphae	$7 \times 10^6 \, g^{-1}$
Length of hyphae	$140 \, m \, g^{-1}$
Volume of hyphae	$0 \cdot 97 \, mm^3 \, g^{-1}$
Number of bacteria and actinomycetes (plate count)	$44 \times 10^6 \, g^{-1}$
Number of bacteria and actinomycetes (direct count)	$1600 \times 10^6 \, g^{-1}$
Fraction of pore space occupied by organisms	$0 \cdot 35 \%$
Microbial biomass from biovolume	$220 \, \mu kg \, C \, g^{-1}$
Microbial biomass from flush (determination A)	$220 \, \mu kg \, C \, g^{-1}$
Microbial biomass from flush (determination B)	$570 \, \mu kg \, C \, ha^{-1}$
ATP content of soil	$1 \cdot 22 \, \mu g \, ATP \, g^{-1}$
Turnover time of biomass carbon	$2 \cdot 5$ years
Maximal value for specific maintenance rate	$0 \cdot 21 \, year^{-1}$
Nitrogen in biomass	$95 \, kg \, N \, ha^{-1}$
Flux of nitrogen through biomass	$38 \, kg \, N \, ha^{-1} \, year^{-1}$
Nitrogen offtake in grain and straw	$24 \, kg \, N \, ha^{-1} \, year^{-1}$
Phosphorus in biomass	$11 \, kg \, P \, ha^{-1}$
Flux of phosphorus through biomass	$4 \cdot 6 \, kg \, P \, ha^{-1} \, year^{-1}$
Phosphorus offtake in grain and straw	$5 \, kg \, P \, ha^{-1} \, year^{-1}$

unfertilized crop, the remainder coming from rainfall and dry deposition. In manured plots, the fluxes are presumably greater, with concomitant increases in plant yield.

4.7   **Micro-organisms in decay**

Classically, the decomposition of organic materials in soil has been considered to be brought about by successional populations of micro-organisms [8], e.g. primary saprophytic colonization is by the sugar fungi, such as *Mucor*, and cellulolytic organisms follow this. However, it is equally relevant to consider non-successional populations which depend on each other and form communities, e.g. the products of one population may stimulate another [21]. Primary species are those that can grow as monocultures on the substrate of interest; in contrast, secondary species cannot metabolize the substrate and rely on metabolites of the primary species and/or lytic products to sustain their growth. The communities can be stable, although it is incorrect to think of them in a steady state. They exhibit oscillations in population sizes of the component organisms and residual, limiting substrate concentration, the amplitude and period of which depend on the environment. It is now clear that such communities can be very stable.

4.8   **Conclusion**

The provision of plant nutrients by the cycling of minerals is one of the most significant activities of micro-organisms in soil. Application of microbial growth principles to those processes might help in understanding the dynamics of the processes such that they could be manipulated. In a first approximation, the growth of micro-organisms using the natural substrates present in soil can be analysed in a similar manner to their growth on laboratory media. However, the chemical complexity of the substrate, its spatial inaccessibility, the slow growth rates attained and the interaction between organisms during the cycling of nutrients all complicate the analysis. This should not be a deterrent and hopefully mathematical modelling will stimulate more informed experimentation.

References

1 Antai S. P. & Crawford D. L. (1981) Degradation of softwood, hardwood, and grass lignocelluloses by two *Streptomyces* strains. *Applied and Environmental Microbiology*, **42**, 378–80.

2 Brown P. L. & Dickey D. D. (1970) Losses of wheat straw residue under simulated field conditions. *Soil Science Society of America Proceedings*, **34**, 118–21.

3 Campbell C. A., Paul E. A., Rennie D. A. & McCallum K. J. (1967) Applicability of the carbon-dating method of analysis to soil humus studies. *Soil Science*, **104**, 217–24.

4 Chapman S. J. & Gray T. R. G. (1981) Endogenous metabolism and macromolecular composition of *Arthrobacter globiformis*. *Soil Biology and Biochemistry*, **13**, 11–18.

5 Cole C. V., Innis G. S. & Stewart N. W. B. (1977) Simulation of phosphorus cycling in a semi-arid grassland. *Ecology*, **58**, 1–15.

6 Dowdell R. J. (1982) Fate of nitrogen applied to agricultural crops with particular reference to denitrification. *Philosophical Transactions of the Royal Society, London B*, **296**, 363–73.

7 Fenchel T. & Blackburn T. H. (1979) *Bacteria and Mineral Cycling*. Academic Press, London.

8 Garrett S. D. (1970) *Pathogenic Root-infecting Fungi*. Cambridge University Press, Cambridge.

9 Harper S. H. T. & Lynch J. M. (1981) The chemical components and decomposition of wheat straw leaves, internodes and nodes. *Journal of the Science of Food and Agriculture*, **32**, 1057–62.

10 Harper S. H. T. & Lynch J. M. (1981) The kinetics of straw decomposition in relation to its potential to produce the phytotoxin acetic acid. *Journal of Soil Science*, **32**, 627–37.

11 Hurst H. M., Burges A. & Latter P. (1962) Some aspects of the biochemistry of humic acid decomposition by fungi. *Phytochemistry*, **1**, 227–31.

12 Jenkinson D. S. & Ayanaba A. (1977) Decomposition of carbon-14 labelled plant material under tropical conditions. *Soil Science Society of America Journal*, **41**, 912–15.

13 Jenkinson D. S. & Ladd J. N. (1981) Microbial biomass in soil measurement and turnover. In *Soil Biochemistry*, vol. 5, eds Paul E. A. & Ladd J. N., pp. 415–71. Marcel Dekker, New York.

14 Lynch J. M. & Panting L. M. (1980) Cultivation and the soil biomass. *Soil Biology and Biochemistry*, **12**, 29–33.

15 Mathur S. P. & Paul E. A. (1967) Microbial utilization of soil humic acids. *Canadian Journal of Microbiology*, **13**, 573–80.

16 Paul E. A. & van Veen J. A. (1978) The use of tracers to determine the dynamic nature of organic matter. *Transactions of the Eleventh International Society of Soil Science Congress (Edmonton)*, pp. 61–102.

17 Paul E. A. & Voroney R. P. (1980) Nutrient and energy flows through soil microbial biomass. In *Contemporary Microbial Ecology*, eds Ellwood D. C., Hedger J. N., Latham M. J., Lynch J. M. & Slater J. H., pp. 215–37. Academic Press, London.

18 Sain P. & Broadbent F. E. (1977) Decomposition of rice straw in soils as affected by some management factors. *Journal of Environmental Quality*, **6**, 96–100.

19 Sauerbeck D. R. & Gonzalez M. A. (1977) Field decomposition of carbon-14 labelled plant residues in various soils of the Federal Republic of Germany and Costa Rica. In *Soil Organic Matter Studies*, vol. 1, pp. 159–70. International Atomic Energy Agency, Vienna.

20 Shields J. A. & Paul E. A. (1973) Decomposition of ^{14}C-labelled plant material under field conditions. *Canadian Journal of Soil Science*, **53**, 297–306.

21 Slater J. H. & Bull A. T. (1978) Interactions between microbial populations. In *Companion to Microbiology*, eds Bull A. T. & Meadow P. M., pp. 181–206. Longman, London.

22 Smith J. H. & Douglas C. L. (1971) Wheat straw decomposition in the field. *Soil Science Society of America Proceedings*, **35**, 269–72.

23 Swift M. J., Heal O. W. & Sanderson J. M. (1979) *Decomposition in Terrestrial Ecosystems*. Blackwell Scientific Publications, Oxford.

THE SPERMOSPHERE AND RHIZOSPHERE: INTERACTIONS OF ASYMBIOTIC MICRO-ORGANISMS AND PLANTS

Deciphering the quaint, fantastic and multitudinous association of micro-organisms that constitute the *corpus delicti* of the rhizosphere region, the author is tempted to hazard the opinion that the tenancy provided by plant roots to alien micro-organisms is amazing and astounding. Root exudates with their tempting constituents, carbon dioxide exhaled by the rootlets searching for crevices for the vital fluids, rejected transient root parts of ephemeral existence such as root hairs, non-cambial feeder rootlets and root cap cells figure conspicuously in the weird and peculiar ecology of the root region.

R. K. Kakkar The romance of the rhizosphere: a new blossoming. *Mycopathologica et Mycologia applicata*, **41**, 347–55, 1970.

From seed germination until the plant reaches maturity, micro-organisms grow associatively unless measures are taken in laboratory experiments to deliberately exclude them. This situation, which is now termed rhizocoenosis, is commonly ignored by plant physiologists and agronomists.

5.1 Anatomy

The region around the germinating seed, the spermosphere, has received little attention from plant physiologists and microbiologists. Yet here, perhaps, there is the greatest potential for microbial effects

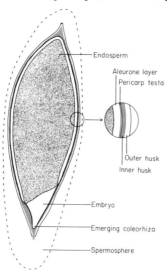

Fig. 5.1 Germinating barley seed and its associated spermosphere. The extent of the spermosphere varies depending on the plant species and soil conditions but it is likely to be greatest in the embryo region.

to have a lasting influence on plants. The influences can affect plant establishment. Sometimes plants will compensate but often the legacy of poor establishment can be seen at harvest. The spermosphere varies between the many seed types; that of a typical cereal is depicted in Fig. 5.1. Some micro-organisms are present under the seed coat; these generally originate at the flowering stage and therefore depend on the air flora at the time. In damp weather, when the fungal flora in the atmosphere is increased, a greater number of species and fungal biomass can colonize the seed during its formation. The colonization can be assessed quantitatively using a lactophenol blue stain [40]. Micro-organisms also colonize the seed coat during storage, especially if the seed moisture content increases much above 16 per cent w/w. The soil is the final inoculum source for seed colonization; some cellulolytic fungi attack the seed coat, while primary saprophytic fungi colonize the embryo region (Fig. 5.2). At the embryo end of some seeds, such as the pea, there is a micropyle which allows oxygen to enter. However, in cereals, there is a 'plug' which

Fig. 5.2 Colonization of a barley seed by *Gliocladium roseum*.

has a similar function [20,26]. The spermosphere flora can include pathogens, e.g. Fig. 5.3 shows the colonization of an asparagus seed coat by *Fusarium*. Several thousand spores can become trapped in natural crevices or insect tunnels.

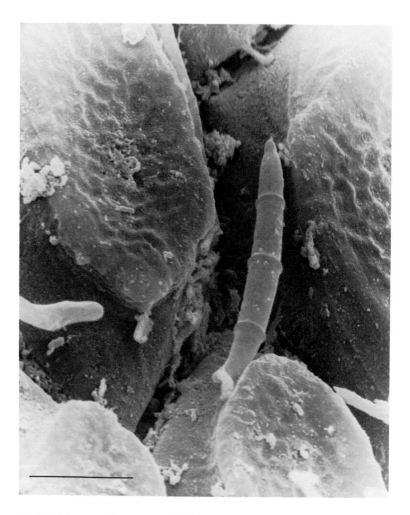

Fig. 5.3 Macroconidium, presumed to be *Fusarium moniliforme*, lodged in a crevice of an asparagus seed coat. Bar marker = 10 μm. (Photograph by D. A. Inglis, Washington State University.)

The rhizosphere was first defined by Hiltner [21] as the zone of stimulated bacterial growth around legumes, resulting from the release of nitrogen compounds by nodules. This description has since been modified and made more general; now it seems reasonable to include all microbial growth using root-derived compounds as sources of carbon, nitrogen and energy. The root epidermis–cortex zone, when colonized by pathogens or non-pathogens [16,32], has been termed the 'endorhizosphere' [3]. It appears, therefore, reasonable to describe the zone of colonization outside the root as the ectorhizosphere [24,25] (Fig. 5.4). However, it must be recognized that the zone where the ecto- and endorhizospheres merge, is probably where some of the most intimate plant and microbial associations occur [19] and this is termed the rhizoplane.

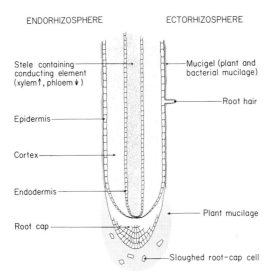

ENDORHIZOSPHERE   ECTORHIZOSPHERE

Stele containing conducting element (xylem↑, phloem↓)

Mucigel (plant and bacterial mucilage)

Root hair

Epidermis

Cortex

Endodermis

Root cap

Plant mucilage

Sloughed root-cap cell

Fig. 5.4   Root region. The ectorhizosphere includes the surrounding soil. Although the endodermis and stele are included in the endorhizosphere, this is normally the specialized primary reserve of pathogens.

The ectorhizosphere and rhizoplane have been viewed by light microscopy, usually after staining with phenolic aniline blue (Fig. 5.5). However, it is not necessary to stain if Nomarski differential interference microscopy is employed, as each cell is very clearly distinguished (Fig. 5.6). The bacteria often occur as distinct colonies, with at least 10^2 cells per colony [12], on the plant root (Fig. 5.5) and are often concentrated at the junctions between epidermal cells (Fig. 5.6). It is just possible, however, that this could be an artifact of the preparation, the bacteria being washed off the convex surfaces of the cortical cells and being retained in the junctions. The non-random distribution of bacteria on the rhizoplane of field-grown plants has been analysed mathematically [27] but it may be more appropriate to use more modern types of statistical analysis, such as those

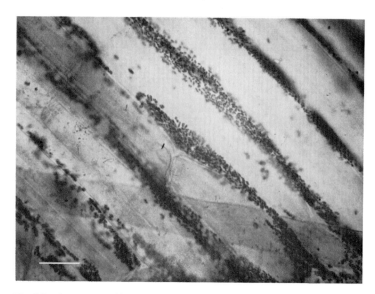

Fig. 5.5   Aggregates of bacteria in the intercellular spaces of the root surface of wheat inoculated with a *Pseudomonas* sp. The population was $c. 7 \times 10^7$ (mg dry wt. root)$^{-1}$. The roots were stained with phenolic aniline blue. Bar marker $= 5\,\mu$m.

Fig. 5.6   Microcolony spreading across the surface of epidermal cells of wheat inoculated as in Fig. 5.5. The roots were mounted unstained in phenolic acetic acid and observed by Nomarski differential interference microscopy. Bar marker $= 5\,\mu$m.

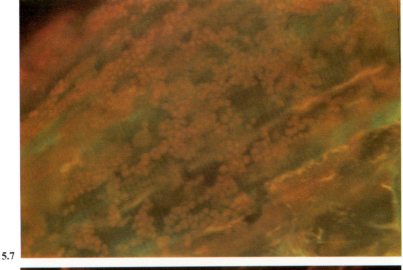

5.7

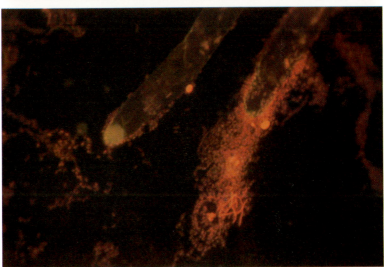

5.8

Fig. 5.7 Short rod-shaped bacteria on the surface of a wheat root grown in agar solution prestained with fluorescent brightener and subsequently stained with europium chelate for 30 min. (Photograph by B. G. Johnen, ICI Plant Protection Division.)

Fig. 5.8 Root hairs and the surrounding micro-organisms of wheat root grown in liquid solution and stained for 30 min with europium chelate and fluorescent brightener formulated in 55% (w/v) ethanol. (Photograph by B. G. Johnen, ICI Plant Protection Division.)

employed for counting bacteria on soil particles [34]. Assessment of the population size is also dependent on all the bacteria adhering to the root surface and in practice the ability to adhere varies greatly between species. Many can be removed along with soil materials during sample preparation.

A useful modification of the light microscopic technique is to stain the micro-organisms with europium chelate and a fluorescent brightener [22] (Figs 5.7 and 5.8). The living microbial cells fluoresce red in ultraviolet light, while dead cells and organic matter fluoresce green. For short exposure times to the stain, the root is not penetrated by the stain and therefore also appears green. Technically, the method is rather difficult to use and requires a good quality microscope, which sometimes limits its routine use. Staining cells with tetrazoliums, which only stain when the dehydrogenases of a cell are

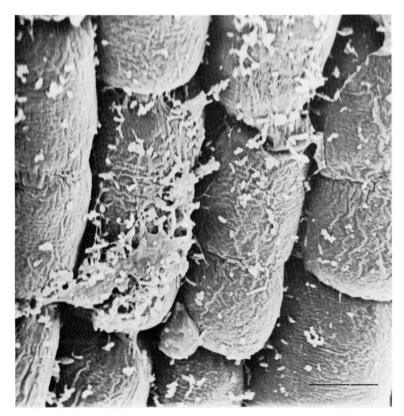

Fig. 5.9   Root surface of maize (grown in solution culture) with bacteria and associated mucigel. Bar marker = 10 μm. (Photograph by R. Campbell, Bristol University and M. C. Drew, Letcombe Laboratory.)

active, has been used to demonstrate the presence of bacteria in the endorhizosphere [33].

The non-uniform colonization patterns of roots in solution culture have also been demonstrated with SEM (Fig. 5.9); it has been used to examine the rhizoplanes of field-grown plants [19] (Figs 5.10 and 5.11). This has shown that the rhizoplane of direct-drilled winter-wheat plants had a greater microbial population than plants drilled into tilled land (Chapter 9). It is more difficult, however, to use SEM rather than light microscopy for the quantitative assessment of microbial populations.

Similarly, whereas TEM of transverse sections has been particularly

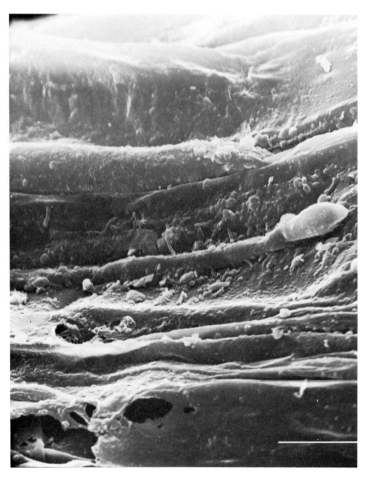

Fig. 5.10   Root surface of wheat (grown in soil) with bacteria and associated mucigel. Bar marker $= 10\,\mu$m. (Photograph by L. F. Elliott, Washington State University.)

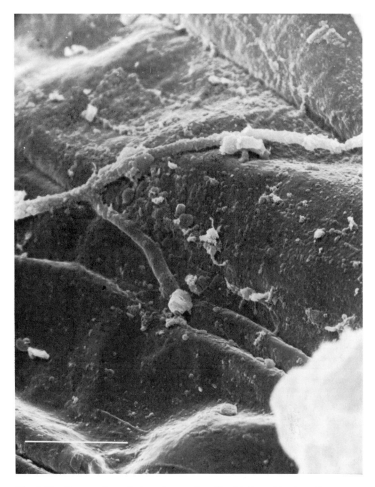

Fig. 5.11   Root surface of wheat (grown in soil) with fungal hyphae. Bar marker = 10 μm. (Photograph by L. F. Elliott, Washington State University.)

useful in depicting the rhizoplane (Fig. 5.12) and endorhizosphere (Fig. 5.13) microflora, the thin sections of root which must be cut severely limit generalizations to be made on population densities and distributions. Even with these limitations, however, all the techniques clearly indicate that there is a much greater biomass in the older parts of the root. The much closer proximity of the endorhizosphere bacteria to the conducting elements (xylem and phloem) of the plant mean that they are spatially better placed to interrupt the plant's nutrient supply and introduce phytotoxins or other plant growth regulators to the translocation stream (Chapter 7). Figure 5.13 shows cortical cell-wall decay by endorhizosphere bacteria

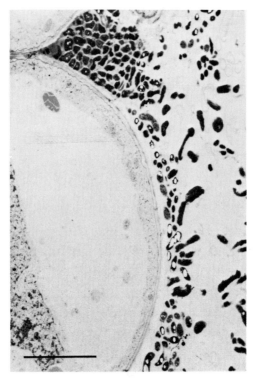

Fig. 5.12 Maize epidermal cell (grown in solution culture) and the attached bacteria. Bar marker = 1 μm. (Photograph by R. Campbell, Bristol University and M. C. Drew, Letcombe Laboratory.)

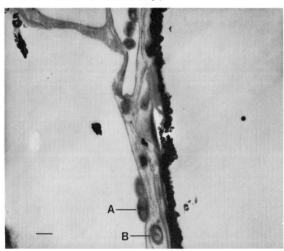

Fig. 5.13 Bacterial cells in a wheat cortical cell (A) and cell wall (B). The plant, nearing maturity, was growing in soil. Bar marker = 1 μm.

embedded in the cell walls. Such cellular decay in crop plants can kill plants or at least reduce yields. However, cortical decay can also be a functional response of the root, e.g. in Fig. 5.14 the decay which results in air spaces, lacunae or aerenchyma is induced by the endogenous production of ethylene (section 7.2) in the root, which itself is induced by anoxia around the roots. The air spaces facilitate the movement of oxygen from shoots to roots [18]. The lysed cells presumably then serve as a substrate for endorhizosphere saprophytes. In some desert plants, rhizosheaths form (Figs 5.15 and 5.16) and the cortex decays (Fig. 5.17). The cause of the cortical decay (Figs 5.15, 5.16 and 5.17) is uncertain. Although there is extensive rhizosphere colonization by roots of helically lobed (*Anacalomicrobium*) and branched (*Hyphomicrobium*) forms and a fungus, *Olpidium*, the process could be autolytic [41]. The decay leaves a sheath around the

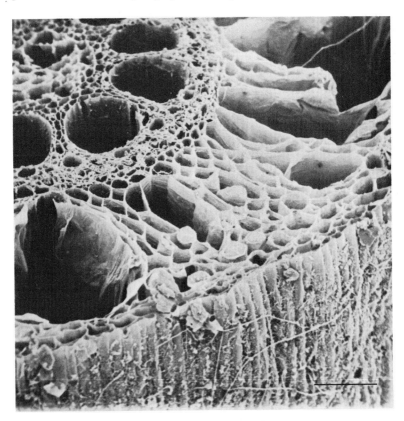

Fig. 5.14   Cross-section of a maize root (grown in solution culture) showing the associated microbial population on the root surface. Bar marker = 100 μm. (Photograph by R. Campbell, Bristol University and M. C. Drew, Letcombe Laboratory.)

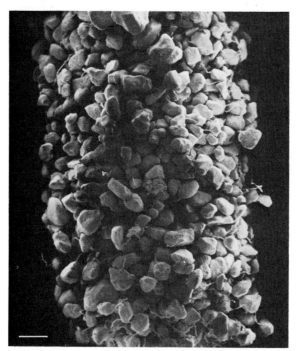

Fig. 5.15 External view of a rhizosheath segment of *Oryzopsis hymenoides* with sand grains attached. Note the root hair tips extending beyond the outer layer of sand grains. Bar marker = 0·2 mm. (Photograph by L. H. Wullstein, University of Utah.)

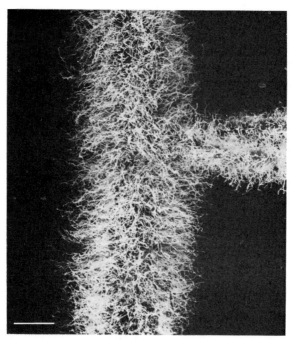

Fig. 5.16 Rhizosheath segment of *Oryzopsis hymenoides* with the sand grains removed to expose the dense mass of root hairs on the primary and lateral roots. Bar marker = 0·5 mm. (Photograph by L. H. Wullstein, University of Utah.)

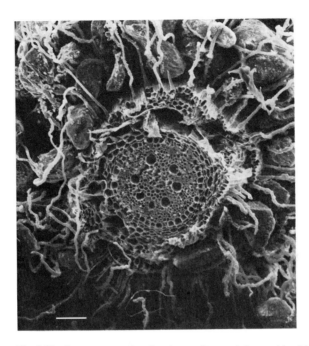

Fig. 5.17   Root cross-section showing an *Oryzopsis hymenoides* rhizosheath matrix of root hairs with attached sand grains. The first signs of lacunae development in the cortex are evident; eventually almost the complete cortex can lyse. Bar marker = 0·1 mm. (Photograph by L. H. Wullstein, University of Utah.)

root. This sheath can be slid from the root. The binding of sand grains to the root hairs may be initiated by the rhizosphere micro-organisms. The function of such sheaths around roots has been discussed [17]. Their contact with the soil is likely to influence the radial movement of ions and water towards root surfaces. Although they might provide some protection against desiccation, the discontinuities in the partly degenerated cortices would be expected to increase resistance to flow and delay the re-establishment of continuous films of water when roots are remoistened after a period of drought.

Microscopic examination has commonly demonstrated that bacteria predominate over fungi in the rhizosphere but it has been shown [30] that in a survey of 40 sites in England and Wales, the fungi/bacteria ratios, expressed as percentage cover of the root surface of *Plantago lancedata*, varied from 0·28 to 14. Using selective antibiotics (streptomycin to inhibit bacteria and actidione to inhibit fungi) it was shown [39] that bacterial respiration was dominant in the rhizosphere but that in the bulk soil fungal respiration dominated. This will depend partly on whether ectomycorrhizas (Chapter 6) develop.

Although the species diversity of the inoculum pool in the soil is great, there are claims that some bacteria, particularly *Pseudomonas* spp., dominate the rhizospheres of particular soils [35]. The reasons for this are unclear and untested. Two possibilities are that adhesion phenomena could take place where extracellular polymers of the root and bacteria interact or that there is chemotaxis, such as that by *Rhizobium* to legumes (section 6.2) or that which *Phytophthora infestans* exhibits to tree roots which produce ethanol in anaerobic soils [1]. An understanding of the mechanisms governing the community structure of the rhizosphere is necessary if it is to be controlled (section 10.4).

5.2   Provision of substrates for microbial growth

During membrane reorganization of the seed embryo, which is the first step in germination, organic materials are released. Thereafter, a slow release of materials may arise from the endosperm (Fig. 5.2). Irrespective of the source, the materials flow from the embryo end of the seed and this accounts for the dominance of microbial colonization in this region.

The organic materials released by roots take various forms. *Exudates* leak from living roots, *secretions* are actively pumped and *lysates* are passively released from roots during autolysis. *Mucigel* arises mainly from *plant mucilage* and is produced from epidermal and root-cap cells. Microbial growth around roots, using exudates, secretions and sloughed cells as substrates, leads to *microbial mucilages* which also form part of the mucigel.

To measure the carbon release in non-sterile soil, plants have been grown with $^{14}CO_2$ as the sole source of carbon available to the plant. The ^{14}C in the plant and the soil and in the carbon dioxide respired by roots and soil micro-organisms was fractionated and hence a carbon balance sheet was produced [9,37]. This showed that about 20 per cent of the dry matter produced by the plants was released by the roots but only about half of this was released into sterile soil. The release of carbon from roots in the presence of bacteria can also be estimated indirectly, by comparing that released in carbon compounds under aseptic conditions with the carbon retained in bacterial cells (*and* provided with no other carbon substrate) growing around roots [8]. This again demonstrated that bacteria appeared to stimulate the release from roots to substrates, the amount of which increased as the root aged.

The precise chemical nature of the carbon compounds released by roots is difficult to determine but, as would be expected, they comprise all the kinds of materials that are found in cells [36].

Carbohydrates are the major components and the carbon/nitrogen ratio is about 30:1 based on the amino acid content [6]. Whereas this is a more balanced substrate than straw, microbial decomposition will still tend to immobilize soil or fertilizer nitrogen.

5.3   **Microbial growth kinetics and mathematical models**

No detailed analysis of microbial growth kinetics in the spermosphere has been undertaken although an approximate method based on staining has shown that as the seed imbibes, then the associated fungal biomass under the husk increases (Fig. 5.18).

The rate at which plants release substrates to the rhizosphere microflora is uncertain but there are gradients of release, with maximal amounts being produced around the zone of cell elongation behind the root tip and where lateral roots are formed [38]. Generally, the size of the rhizosphere population per unit weight of root increases with the growth and age of the roots (Fig. 5.19). The stimulation is somewhat analogous to continuous culture (section 4) and a mathematical analysis of the substrate flow has been produced [29]. The change in substrate concentration ($\mu g\,cm^{-3}$ soil), S, with time, t, in the rhizosphere/rhizoplane region was analysed, considering the root as an idealized cylinder, and expressed as:

$$\frac{dS}{dt} = F_D + F_I - (F_G + F_M).$$

| F_D (substrate to a radial distance, r, from root axis) | F_I (indigenous supply of substrate from soil) | F_G (substrate used in microbial growth) | F_M (substrate used in microbial maintenance) |

The model was tested to assess the influence of some individual variables of the limited amount of data tested with the model and reasonably good fits were obtained. However, because of the lack of experimental data, the same restrictions apply as for the other models of growth in the soil (Chapter 4).

A chemostat has been used as an experimental model for the rhizosphere [15]. The substrate was glucose (to mimic root-derived carbon). A mixed population of rhizosphere and associated organisms was inoculated: *Pseudomonas* sp. (a true rhizosphere inhabitant), *Arthrobacter* sp. (an autochthonous soil bacterium [see section 3.1]), *Acanthamoeba* sp. (a protozoan predator of bacteria) and *Mesodiplogaster* sp. (a nematode predator of protozoa and bacteria). Some interesting population dynamics were observed but the substrate and community were unreal and it would have been more appropriate to use a complex substrate with a whole community (section 4) of organisms actually isolated from the rhizosphere.

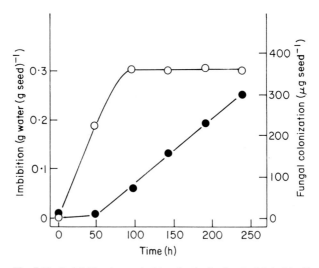

Fig. 5.18 Imbibition (open circle) and colonization (solid circle) of barley seed incubated at high relative humidity. (From Harper & Lynch [20].)

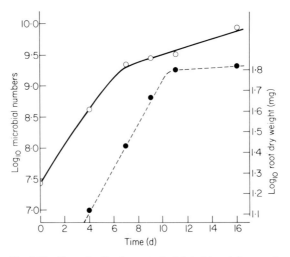

Fig. 5.19 Growth of barley roots (solid circle) and the associated population of bacteria (open circle). (From Barber & Lynch [8].)

Although the chemostat provides one model, rhizosphere population dynamics can be examined in conditions closer to nature by growing plants under gnotobiotic conditions with different microorganisms. Bowen [4] has argued eloquently in favour of such an approach and suggested that the propagules of the rhizosphere community, especially fungi, can be divided into *r*- and *K*-strategists, a concept used for higher plant and animal ecology (Table 5.1). The growth of a *Curtobacterium* sp. (isolated from the barley rhizosphere), *Mycoplana* sp. (isolated from oil-seed rape) and *Pseudomonas* sp. (isolated from maize) in the wheat rhizosphere has been studied [12]. When inoculated individually they all reached a similar population size, the *colonization potential*, irrespective of inoculum size [13] (Fig. 5.20). When inoculated together the *Pseudomonas* sp. again reached a similar population size but the *Curtobacterium* sp. was unable to compete (Fig. 5.21). Such basic studies could lead to a better understanding of the potential efficiency of seed inoculants

Table 5.1 Population strategies of rhizosphere propagules. (After Bowen [14].)

r-strategy	*K*-strategy
Small propagules	Large propagules
Opportunistic	High competitive ability
Rapid growth rates	Longevity
High mortality of population but rapid recovery	Low mortality rate
	Slow recovery from population reduction
Rapid germination and growth on simple substrates	Late colonizers of rhizosphere
Early colonizer of rhizosphere	Grow over larger distances

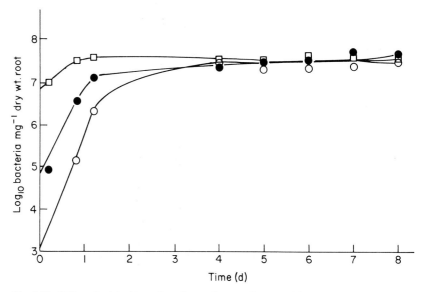

Fig. 5.20 Effect of original inoculum size on the growth rate and final population of a fluorescent *Pseudomonas* sp. on barley roots (mean of counts from the middle and older portions of the root). (After Bennett & Lynch [13].)

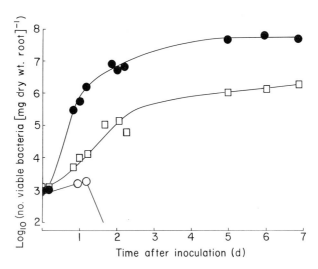

Fig. 5.21   Growth of a *Curtobacterium* sp. (open circle), a *Pseudomonas* sp. (solid circle) and a *Mycoplana* sp. (open square) when co-inoculated in the wheat rhizosphere. (From Bennett & Lynch [12].)

(section 10.4). Studies made on the growth of *Rhizobium legumino-sarum* and *Arthrobacter crystallopoietes*, which were streptomycin resistant, in the rhizoplane of soil-grown barley plants showed that the nodule bacterium stabilized at the same abundance regardless of the initial population density, whereas the abundance of *A. crystallopoietes* depended on the inoculum size [23]. In this respect *R. leguminosarum* was inclined to an *r*-strategy and *A. crystallopoietes* appeared as a *K*-strategist.

5.4   **Microbial effects on ion uptake by plant roots**

Bacteria influence the uptake of ions by plant roots growing under gnotobiotic conditions [4]. Under some conditions, ion uptake by plant roots can be stimulated by bacteria, possibly by providing chelating agents or plant growth regulators (section 7.2) to promote active ion transport. Conversely, under other conditions they can be inhibitory, either by competing for the nutrients or producing phytotoxic compounds (section 7.3). Phosphate, which occurs in the soil solution in small concentrations and is readily incorporated by bacteria into nucleic acids, has received the greatest attention. Using ^{32}P-labelled phosphate, it can be seen that the presence of bacteria can cause phosphate accumulation on the rhizoplane [10] (Fig. 5.22). Over short (30 min) periods phosphate uptake and translocation within the plant are promoted by bacteria but over longer (24 h)

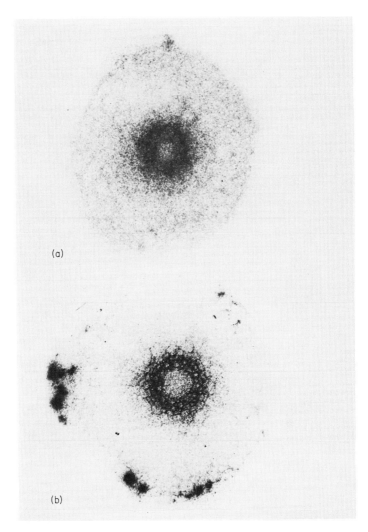

Fig. 5.22   Transverse sections of barley roots in solution culture showing the uptake of ³²P-labelled phosphate: (a) sterile and (b) with bacteria present. (Photograph by D. A. Barber, Letcombe Laboratory.)

periods uptake is reduced [5]. Seedling age is also important, phosphate uptake being stimulated by the bacteria in younger (6 d) seedlings and inhibited in older (12 d) seedlings. Bacteria appear to promote the uptake of manganese (Table 5.2) by producing some type of chelating agent [7].

Few experiments on ion uptake have been done in soil. In gamma-radiation-sterilized soil, the phosphorus content of plants can be greater [11] but this might be an indirect chemical effect of the

Table 5.2 Effect of micro-organisms on the uptake over 24h of labelled manganese $(0.02 \, mol \, m^{-3})$ by three-week-old barley plants. (After Barber & Lee [9].)

Origin of solution for uptake studies	Condition of plants	Uptake of Mn $(g \, g^{-1} \, dry \, wt.)$	
		Root	Shoot
Sterile plants	Sterile	248	8.6
Non-sterile plants	Sterile	506	12.7
Non-sterile plants	Non-sterile	1366	21.5

irradiation. Such difficulties in interpretation can frequently occur. Certainly phosphate-solubilizing bacteria occur in the rhizosphere. In one study their numbers were greater in the cowpea than in the maize rhizosphere [31] but the significance of this in terms of phosphorus availability to the plants is uncertain.

In the rice rhizosphere, *Bacillus* sp. and *Pseudomonas* sp. appear to dominate [2]. The *Pseudomonas* sp., unlike the *Bacillus* sp., can accumulate nitrite from nitrate in pure culture. As the inhibitory effect of *Pseudomonas* sp. on rice was similar to the application of nitrite, it was concluded that these organisms inhibit plant growth by providing nitrite.

Glasshouse crops, particularly tomatoes, are now often grown in continuously recirculating water culture where there is a shallow layer of nutrient solution circulating between two layers of polyethylene sheet. This is known as the nutrient-film technique and gives control of the root environment. It removes the need for watering. Uniformity in the supply of nutrient and crop protection chemicals is ensured, root temperature can be raised when necessary and the soil problems of salinity, poor structure and drainage are avoided. Hitherto, pathogens have only rarely been a problem. However, it would appear that the studies of the role of micro-organisms in ion uptake from nutrient solution are particularly relevant to this technique and should be considered in management.

5.5   Interactions of the rhizosphere of one plant with that of another

The abundance of root-surface fungi and bacteria and internal mycorrhizae were assessed in three grass species and a clover grown alone and in two-species mixtures [28]. The abundance depended on the plant nitrogen concentration but in glasshouse and field experiments the abundance of all three microbial groups was sometimes significantly affected by the presence of another species of plant. Often the fungal abundance was intermediate between two monoculture values, suggesting that the fungi were using substrates from both plants. There was some evidence that mycorrhizal development on white clover (*Trifolium repens*) could be inhibited by the presence of sweet vernal grass (*Anthoxanthum odoratum*) and it was suggested that this might be due to the release of a toxic root exudate. Thus, studies could have relevance for the selection of species when reseeding grassland (section 9.5) and in the increasingly common practice of mixed cropping of legumes with grasses or cereals. The interaction between plant species is usually considered as allelopathy and the chemical basis for this is discussed further in Chapter 7.

5.6   **Conclusion**

Much of the management of plant–micro-organism interactions in the soil and in recirculating water culture depends on understanding and controlling the spermosphere and rhizosphere populations. Such management potentials are discussed more fully in Chapter 10. Whereas these applications are commendable targets, they must be based on sound scientific foundation and this will involve fundamental studies under gnotobiotic conditions as well as tests under field conditions.

References

1 Allen R. N. & Newhook F. J. (1973) Chemotaxis of zoospores of *Phytophthora cinnamomi* to ethanol in capillaries of soil pore dimensions. *Transactions of the British Mycological Society*, **61**, 287–302.

2 Asanuma S., Tanaka H. & Yatazawa M. (1980) *Pseudomonas capacia*—A characteristic rhizoplane micro-organism in rice plant. *Soil Science and Plant Nutrition*, **26**, 71–8.

3 Balandreau J. & Knowles R. (1978) The rhizosphere. In *Interactions Between Non-pathogenic Soil Micro-organisms and Plants*, eds Dommergues Y. R. & Krupa S. V., pp. 243–68. Elsevier, Amsterdam.

4 Barber D. A. (1978) Nutrient uptake. In *Interactions Between Non-pathogenic Soil Micro-organisms and Plants*, eds Dommergues Y. R. and Krupa S. V., pp. 131–62. Elsevier, Amsterdam.

5 Barber D. A., Bowen G. D. & Rovira A. D. (1976) Effects of micro-organisms on the absorption and distribution of phosphate in barley. *Australian Journal of Plant Physiology*, **3**, 801–8.

6 Barber D. A. & Gunn K. B. (1974) The effect of mechanical forces on the exudation of organic substances by the roots of cereal plants grown under sterile conditions. *New Phytologist*, **73**, 39–45.

7 Barber D. A. & Lee R. B. (1974) The effect of micro-organisms on the absorption of manganese by plants. *New Phytologist*, **73**, 47–106.

8 Barber D. A. & Lynch J. M. (1977) Microbial growth in the rhizosphere. *Soil Biology and Biochemistry*, **9**, 305–8.

9 Barber D. A. & Martin J. K. (1976) The release of organic substances by cereal roots in soil. *New Phytologist*, **76**, 69–80.

10 Barber D. A., Sanderson J. & Russell R. S. (1968) Influence of micro-organisms on the distribution in roots of phosphate labelled with phosphorus-32. *Nature (London)*, **217,** 644.

11 Benians G. J. & Barber D. A. (1974) The uptake of phosphate barley plants from soil under aseptic and non-sterile conditions. *Soil Biology and Biochemistry*, **6,** 195–200.

12 Bennett R. A. & Lynch J. M. (1981) Bacterial growth and development in the rhizosphere of gnotobiotic cereal plants. *Journal of General Microbiology*, **125,** 95–102.

13 Bennett R. A. & Lynch J. M. (1981) Colonization potential of rhizosphere bacteria. *Current Microbiology*, **6,** 137–8.

14 Bowen G. D. (1980) Misconceptions, concepts and approaches in rhizosphere biology. In *Contemporary Microbial Ecology*, eds Ellwood D. C., Hedger J. N., Latham M. J., Lynch J. M. & Slater J. H., pp. 283–304. Academic Press, London.

15 Coleman D. C., Cole C. V., Hunt H. W. & Klein D. A. (1978) Trophic interactions in soils as they affect energy and nutrient dynamics. I. Introduction. *Microbial Ecology*, **4,** 345–9.

16 Darbyshire J. F. & Greaves M. P. (1973) Bacteria and protozoa in the rhizosphere. *Pesticide Science*, **4,** 349–60.

17 Drew M. C. (1979) Root development and activities. In *Arid-land Ecosystems: Structure Functioning and Management*, vol. 1, eds Perry R. A. & Goodall D. W., pp. 573–606. Cambridge University Press, Cambridge.

18 Drew M. C., Jackson M. B. & Giffard S. (1979) Ethylene-promoted adventitious rooting and development of cortical air spaces (aerenchyma) in roots may be an adaptive response to flooding in *Zea mays* L. *Planta*, **147,** 83–8.

19 Elliott L. F., Gilmour C. M., Lynch J. M. & Tittemore D. (1983) Bacterial colonization of plant roots. In *Microbial–Plant Interactions*, ed. Todd R. L. Soil Science Society of America, Madison. (In press.)

20 Harper S. H. T. & Lynch J. M. (1981) Effects of fungi on barley seed germination. *Journal of General Microbiology*, **122,** 55–60.

21 Hiltner L. (1904) Uber neuere Erfahrungen und Probleme auf dem Gebiet der Bodenbakteriologie und unter besonderer Berucksichtigung der Frundungung und Brache. *Arbeiten der Deutschen Landwirtschaftsgesellschaft Berlin*, **98,** 59–78.

22 Johnen B. G. (1978) Rhizosphere micro-organisms and roots stained with europium chelate and fluorescent brightener. *Soil Biology and Biochemistry*, **10,** 495–502.

23 Kirillova N. P., Stasevich G. A., Kozhevin P. A. & Zvyagintsev D. G. (1981) Bacterial population dynamics in a soil–plant system. *Microbiology*, **50,** 94–8.

24 Lynch J. M. (1981) Interactions between bacteria and plants in the root environment. In *Bacteria and Plants*, eds Rhodes-Roberts M. E. & Skinner F. A., pp. 1–23. Academic Press, London.

25 Lynch J. M. (1982) The rhizosphere. In *Experimental Microbial Ecology*, eds Burns R. G. & Slater S. H., pp. 1–23. Blackwell Scientific Publications, Oxford.

26 Lynch J. M. & Pryn S. J. (1977) Interaction between a soil fungus and barley seed. *Journal of General Microbiology*, **103,** 193–6.

27 Newman E. I. & Bowen H. J. (1974) Pattern of distribution of bacteria on root surfaces. *Soil Biology and Biochemistry*, **6,** 205–9.

28 Newman E. I., Campbell R., Christie P., Heap A. J. & Lawley R. A. (1979) Root micro-organisms in mixtures and monocultures of grassland plants. In *The Soil-Root Interface*, eds Harley J. L. & Russell R. S., pp. 161–73. Academic Press, London.

29 Newman E. I. & Watson A. (1977) Microbial abundance in the rhizosphere: a computer model. *Plant and Soil*, **48**, 17–56.

30 Newman E. I., Heap A. J. & Lawley R. A. (1981) Abundance of mycorrhizas and root-surface micro-organisms of *Plantago lanceolata* in relation to soil and vegetation: a multivariate approach. *New Phytologist*, **89**, 95–108.

31 Odunfa V. S. A. & Oso B. A. (1978) Bacterial populations in the rhizosphere soils of cowpea and sorghum. *Revue d'Ecologie et de Biologie du Sol*, **15**, 413–20.

32 Old K. M. & Nicolson T. H. (1978) The root cortex as part of a microbial continuum. In *Microbial Ecology*, eds Loutit M. W. & Miles J. A. R., pp. 291–4. Springer-Verlag, Berlin.

33 Patriquin D. G. & Dobereiner J. (1978) Light microscopy observations of tetrazolium-reducing bacteria in the endorhizosphere of maize and other grasses in Brazil. *Canadian Journal of Microbiology*, **24**, 734–42.

34 Polonenko D. R., Pike D. J. & Mayfield C. I. (1978) A method for the analysis of growth patterns of micro-organisms in soil. *Canadian Journal of Microbiology*, **24**, 1262–71.

35 Rouatt J. W. & Katznelson H. (1961) A study of the bacteria on the root surface and the rhizosphere soil of crop plants. *Journal of Applied Bacteriology*, **24**, 164–71.

36 Rovira A. D. (1965) Plant root exudates and their influence upon soil micro-organisms. In *Ecology of Soil-borne Plant Pathogens—Prelude to Biological Control*, eds Baker K. F. & Snyder W. C., pp. 170–86. John Murray, London.

37 Sauerbeck D. R. & Johnen B. G. (1977) Root formation and decomposition during plant growth. In *Soil Organic Matter Studies*, vol. 1, pp. 141–8. International Atomic Energy Agency, Vienna.

38 Schippers B. & van Vuurde J. W. L. (1978) Studies of microbial colonization of wheat roots and the manipulation of the rhizosphere microflora. In *Microbial Ecology*, eds Loutit M. W. & Miles J. A. R., pp. 295–8. Springer-Verlag, Berlin.

39 Vancura V. & Kunc F. (1977) The effect of streptomycin and actidione on respiration in the rhizosphere and non-rhizosphere soil. *Zentralblatt fur Bakteriologie, Parasitenkunde, Infektionskrankheiten und Hygiene* (Abt. 2), **132**, 472–8.

40 Warnock D. W. (1971) Assay of fungal mycelium in grains of barley, including the use of the fluorescent antibody technique for individual fungal species. *Journal of General Microbiology*, **67**, 197–205.

41 Wullstein L. H. & Pratt S. A. (1981) Scanning electron microscopy of rhizosheaths of *Oryzopsis hymenoides*. *American Journal of Botany*, **68**, 408–19.

INFECTIONS OF PLANT ROOTS:
SYMBIOTES AND PATHOGENS

The summer's flower is to the summer sweet,
Though to itself it only live and die,
But if that flower with base infection meet,
The basest weed outbraves his dignity.

W. Shakespeare Sonnet XCIV, 1609.

The major beneficial symbiotic associations with roots of agricultural crops are those of the root-nodule bacteria with legumes and the mycorrhiza which associate with most crops. On account of their potential economic significance, these and the root pathogens have received more attention than most other soil micro-organisms. The following is only a brief introduction to the vast literature on these subjects and the reader should consult more detailed texts [1,6,8, 14,24,28,31,33,35].

Whereas the study of the root-nodule bacteria and mycorrhizal fungi has been traditionally considered the preserve of the soil micro-biologist, pathogens have warranted a separate discipline—plant pathology. In many ways, this distinction is quite artificial as both microbiologists and pathologists are interested in the activities of micro-organisms which influence plants. The barriers to communication between the two groups, however, are evident at all levels, including international congresses, scientific societies and university departments. In the section on pathogens (section 6.4), a few pathology concepts are given for the microbiologist to initiate a dialogue with the pathologist.

6.1 Types of interaction

Many parasites are pathogens and vice versa but there are many exceptions to this, e.g. some pathogens produce toxins without parasitizing the plant, whereas mycorrhiza parasitize the plant with no obvious adverse effect. What then are plant symbiotes and pathogens and indeed how do organisms generally interact with each other?

Starr [34] has indicated that, 'The terminology presently used for labelling organismic associations is confusing, parochial, and highly imprecise'. Few people would seriously question this statement and Starr [34] and Lewis [25] have produced schemes which attempt to remedy the situation. The following is a blend of their ideas. It seems that most associations can be described as *symbioses* because this should merely imply a living together in an intimate association and not the special situation where the association is mutually beneficial. Instead of using just the classical terminology, which is often loose in meaning, all associations between organisms can be classified into continua (a philosophical term for an unbroken course of sensations and events). The number of continua which should be used is open to question but the following appear to satisfy most needs.

1. *Locational–occupational.* This continuum describes the spatial position of the association and the means by which it is achieved. The relative sizes of the organisms and their degree of integration can also be studied under this heading, although it could be argued that each aspect of this continuum should be considered as a separate entity. This is perhaps the fundamental continuum and the following classical terms (with their dictionary definitions which may be difficult to understand) should be considered here: *commensal* (messmate which cannot survive without a partner; received at the table of a neighbour without living on it); *parasite* (lives on or in the body of the host; feeds on the cells, tissues or body fluids of another organism; metabolically dependent on the host; adversely affects the other; requires some vital factor from another organism; uses other living organisms as environment and source of food; involves overt exploitation); and *predator* (feeds on; causes death of; consumes; destroys; eats; pursues; hunts; pounces on; kills; captures; uses as food; seizes; exploits). In practice it is often difficult to distinguish between parasitism and predation, although 'predator' usually implies movement.

2. *Valuational.* The benefit or harm of the association is considered in this continuum. The terminology which would normally be used is *mutualism* (both partners gaining benefit), *neutralism* (neither partner affecting the other) and *antagonism* (one of the partners suffering). *Synergism* is a particular type of mutualism in which the formation of specific products is greater in mixed than in pure populations. An *ammensal* relationship concerns the repression of one species by the toxic products of another.

3. *Dependency.* The subject of this continuum is whether or not the association is necessary to each of the partners. The mode of the dependence (chemical, physical and biological) must also be considered. The dependence is either *obligate* or *facultative.*

4. *Durability.* This is a temporal criterion and the association is *persistent* or *transient.*

5. *Specificity.* Various degrees are involved, ranging from highly *specific* (limited to particular associations) or essentially non-specific (*permissive*).

6. *Nutritional.* The major division here is between *autotrophy* and *heterotrophy* (section 3.1). Heterotrophy is of great significance to organism interactions and needs to be subdivided in order to be more specific (Fig. 6.1). Emphasis must be placed on the broken lines in Fig. 6.1 as they show that the placement of organisms into specific

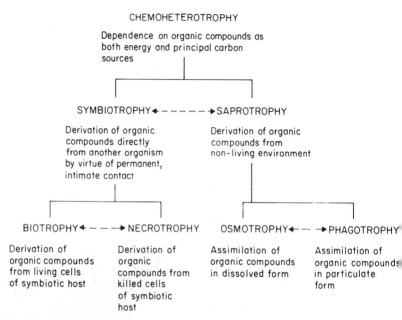

Fig. 6.1 Patterns of chemoheterotrophic nutrition. The broken lines indicate that organisms may behave facultatively in either of the arrowed modes. (After Lewis [25].)

divisions often depends on the habitat and environment. It should also be recognized that the second dichotomy does not produce four equivalent categories; *osmotrophy* and *phagotrophy* concern mechanisms of assimilating nutrient, whereas *biotrophy* and *necrotrophy* are also concerned with the source of nutrients.

The continua as detailed should not be regarded as mutually exclusive and indeed there are many other ways in which they could be formulated, such as basing them on ecological function only. However, an analysis of symbiotes and pathogens at this level aids evaluation of their function. Essentially, the aim should be to describe in simple terms what is seen in the association and not just to fit a single term with ambiguous meaning.

6.2   Root-nodule bacteria

Using the scheme for the description of interactions, the root-nodule bacteria are: (1) within the root by forming infection threads; (2) beneficial to both partners (nitrogen to the plant; carbon to bacteria); (3) facultatively dependent; (4) persistent during the growth of the plant; (5) highly specific to plant and bacterial species; and (6) biotropic.

Some woody dicotyledonous plants have root nodules with acti-nomycetes (*Frankia* spp.) but these have been studied less than the legume root nodules. The techniques of immunofluorescence [4] and intrinsic antibiotic resistance [3], which were described in Chapter 3, have been particularly applied to the study of *Rhizobium* spp., the root-nodule bacteria in soil. Figure 6.2 shows a nodule preparation

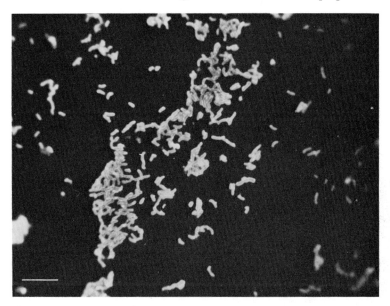

Fig. 6.2 Soybean nodule squash. The strain *Rhizobium japonicum* 61A72 visualized by immunofluorescence (producing a green colour on a dark background). Bar marker = 10 μm. (Photograph by E. L. Schmidt, University of Minnesota.)

viewed by immunofluorescence. One problem is that root-nodule bacteria recovery from soil can be difficult because they attach firmly to clay particles [4] and there is also no selective medium. They are a vitally important source of nitrogen for legumes in the absence of fertilizer nitrogen and therefore reduce the need for expensive nitro-gen fertilizers. However, the addition of fertilizer nitrogen can some-times reduce nodule initiation and/or nitrogenase function. It has been estimated [6] that there were 250×10^6 ha of legumes in the world, providing about $140 \, \text{kg N ha}^{-1} \, \text{year}^{-1}$.

Rhizobium species infect specific hosts, some growing faster than others (Table 6.1). Where the specifity problem of recognition is overcome, such as with *Phaseolus vulgaris* and *R. leguminosarum*, normal nodules are formed but no dinitrogen fixation occurs. There can be great variation in infectivity and effectiveness between strains.

Table 6.1   Host specificity of root-nodule bacteria.

Host	*Rhizobium* spp.	
Alfalfa group	*R. meliloti*	
Clover	*R. trifolii*	Fast
Pea	*R. leguminosarum*	growing
Bean	*R. phaseoli*	
Lupin	*R. lupini*	Slow
Soybean	*R. japonicum*	growing

In most countries, farmers inoculate seeds with effective strains. Usually, the inoculant is applied with a peat or clay base. However, in the UK it is considered that the indigenous strains of *Rhizobium* are adequate and only alfalfa is inoculated because that is not commonly grown and the indigenous strains have not built up a sufficiently large population density.

It was previously thought that *Rhizobium* was an obligate symbiote in fixing dinitrogen but it has now been shown that it grows and fixes it on a defined medium in the laboratory. It has also been shown recently that it will denitrify to N_2O as well as fix dinitrogen [38]. Denitrification requires anaerobic conditions but dinitrogen fixation needs micro-aerobic (low oxygen) conditions. When within the plant root, it is unlikely that anoxia will develop unless the soil becomes waterlogged. Indeed, *Rhizobium* does not appear effective under anaerobic soil conditions [33].

Initiation of nodules might result from the chemotaxis of the bacterium to carbon compounds released by plant roots, particularly at certain points on the root hair. The positive chemotaxis of *R. leguminosarum* towards the root exudates of its host, the edible pea, has been demonstrated [16]. Cationic, neutral and anionic fractions, with molecular weights below 1000, were all attractants and these included amino acids, sugars and carboxylic acids. However, other *Rhizobium* spp. and *Escherichia coli*, which does not grow around roots, were also attracted by pea exudate and *R. leguminosarum* and the other bacteria were attracted by exudates from roots of a range of plants, including non-legumes. Although positive chemotaxis may facilitate infection by *Rhizobium*, it may have little or no role in host–symbiont specificity. However, *E. coli*, unlike *Rhizobium*, exhibits negative taxes to some of the exudate constituents and this may be more ecologically significant than the positive taxes.

It has been argued [12] that specific plant lectins (proteins or glycoproteins) are involved in the initial bonding process of *Rhizobium* to roots, thereby binding to carbohydrates of the bacterial cell wall, e.g. D-mannose, D-glucose, D-fructose and L-sorbose are bonded to pea lectin. The bonding process is very specific and may explain the bacterium–host specificity. However, any excitement over this idea should be tempered because some investigators have not been able to repeat the work.

The bacteria cause curling of the root hair tip and form an infection thread, possibly by invagination (Fig. 6.3). The thread penetrates the cortex and then induces the formation of meristems. The bacteria multiply and are eventually released into the cytoplasm, each surrounded by a membrane. In most temperate (but not tropical) species

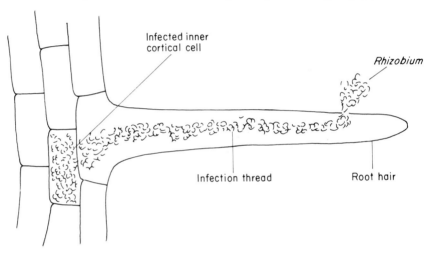

Fig. 6.3 Infection of a root hair by *Rhizobium*. The infected inner cortical cell swells and divides to initiate the nodule.

the bacteria become enlarged. Leghaemoglobin, a protein produced by the plant, controls oxygen diffusion in such a way that the necessary micro-aerobic conditions for dinitrogen fixation are maintained. Thus, an adequate flux of oxygen reaches the rhizobia at the correct tension to allow oxidative phosphorylation without inactivating the nitrogenase enzyme. The bacteria change shape into bacteroids (Fig. 6.4) and no longer multiply. This means that no substrate energy is necessary for growth and more can be diverted to dinitrogen fixation. It is also important that the rhizobia lose their ability to assimilate fixed nitrogen. However, the nitrogenase enzyme needs a similar amount of carbohydrate for its function as the free-living dinitrogen fixer.

The genetics, biochemistry and physiology of dinitrogen fixation by both symbiotes and non-symbiotes have now been studied

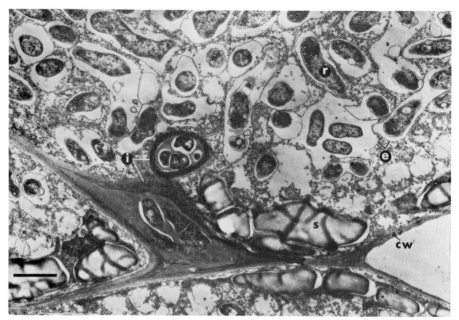

6.4a

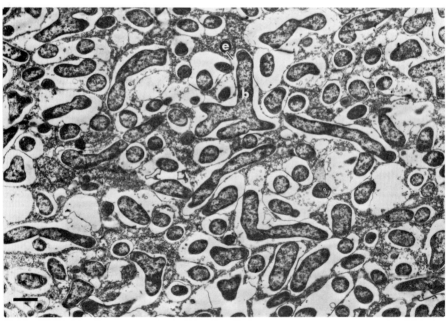

6.4b

Fig. 6.4 (a) Pea nodule showing the infection threads (i) containing bacteria (rhizobia). In the young invaded cell, the rhizobia (r) are still rod shaped. The young cells contain starch (s). The cell wall (cw) and membrane envelope (e) are also apparent. (b) Mature cell showing nitrogen-fixing bacteroids (b) each within its own membrane envelope (e). These cells contain little or no starch. Bar markers = 1 μm. (Photographs by Muriel Chandler, Rothamsted Experimental Station.)

extensively. Two principal objectives of such studies have been to either produce an enzymic process to rival the Haber-Bosch process for producing nitrogen fertilizer or to induce dinitrogen fixation for non-legumes by transferring the nitrogen-fixing (*nif*) genes to the plant directly or through a microbe–plant association. At present, some success has been achieved by showing that the *nif* genes can be transferred to a yeast [39] (Fig. 6.5). *Klebsiella pneumoniae* has 17 *nif*

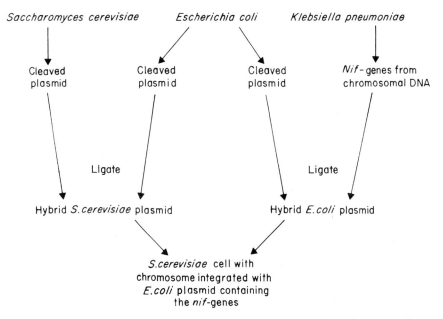

Fig. 6.5   Introduction of the genes for nitrogen fixation (*nif*) into the genome of a yeast. (After Zamir [39].)

genes carried on a chromosome. The genetic elements were transferred to yeast using *E. coli*, the microbiologist's 'genetic workhorse'. Plasmids (loops of DNA separate from the bacterial chromosome) from *E. coli* were cleaved and then fused together to form hybrid plasmids.

The inefficiency of nitrogenase in all dinitrogen fixers is in part caused by the wasteful release of hydrogen gas as a by-product in the enzymatic conversion of dinitrogen to ammonia. In at least one strain of *Rhizobium*, *hup*, a gene or genes carried on a plasmid, code for the synthesis of the enzyme hydrogenase, which breaks down the hydrogen into protons and electrons for reapplication as energy sources in nitrogenase. Introduction of *hup* into *Rhizobium* increased soybean seed yields as compared with yields by plants infected with strains which did not synthesize the hydrogenase [7,13].

6.3   **Mycorrhizal relationships**

Most plants form non-pathogenic associations between their roots and fungi (mycorrhiza); the sedges, crucifers, some chenopodiaceae (e.g. sugar-beet) and certain aquatics are the exceptions. However, the resistance to invasion is probably created by glucosinolates and even crucifers that are unable to form these are susceptible to invasion [20].

There are three major types of mycorrhiza [18,19,26]:

1. *Sheathing or ectomycorrhiza.*   The sheath is a thick, fungal mantle around the root (Fig. 6.6). The Hartig net is within the cortical region

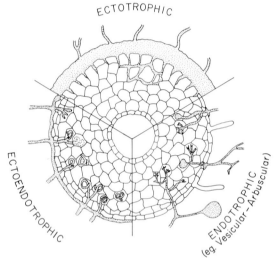

Fig. 6.6   Mycorrhiza types. (a) Ectotrophic. There is a sheath of fungal tissue and limited intercellular invasion by hyphae (the Hartig net). (b) Endotrophic (vesicular–arbuscular as the example). Root hair formation is not suppressed and hyphae ramify surrounding soil producing large spores. (c) Ectoendotrophic. Some of the coiled, fungal hyphae within the plant cells become digested.

of the root, formed by intercellular fungal hyphae. These fungal associations are particularly important in forest trees [18] and are responsible for dichotomous root branching but much less attention has been given to them in crop plants. Like all groups of mycorrhiza, a major function in phosphorus-deficient soils is the transport of phosphorus to the root from the soil. There is controversy as to whether they provide auxin (Chapter 7) to the root but they do appear to enhance plant growth. A wide range of fungi enter into these associations, including basidiomycetes and fungi imperfecti.

2. *Ericaceous or ectendotrophic.*   These are found in the fine rootlets of Ericaceae such as *Caluna* (heather). The intracellular hyphae are

lysed and digested to provide a source of carbon and phosphorus to the host plant.

3. *Non-sheathing or orchidacous/vesicular arbuscular (v–a) or endo-trophic.* All members of the orchid family have these associations but they are now recognized very commonly in most agricultural species, except the Cruciferae. The lower fungi, *Rhizoctonia* and Basidiomycetes, are associated with orchids. Crop species have Zyg-omycetes, particularly *Glomus*, but now other genera are recognized. However, the classification is difficult and confused, especially as no v–a mycorrhiza has yet been grown in culture but it must surely only be a matter of time before the necessary growth factors and nutrients or inhibitors are determined.

Recently, a novel scheme has been proposed to classify mycor-rhizas. It identifies interrelationships between various kinds of mycorrhizas and some of the features which link and distinguish them (Fig. 6.7).

The arbuscles of v–a mycorrhiza (Fig. 6.8) within the plant act as nutrient transfer sites to release carbon and phosphorus and their major function appears to be the provision of the latter in phosphorus-deficient soils. Infection does not occur readily where phosphorus is abundant in soil. Clearly, there is a great prospect for soil inoculation when it is deficient, such as in the hill country of New Zealand [30], and in the reclamation of colliery spoils. Legumes are particularly beneficial because they can have both mycorrhiza and nodules.

The process of infection has been described by Tinker [37] as follows:

$$\frac{dL_i}{dt} = nAe^{Rt} - L_i = SL_i\,(nL_t - L_i) \tag{6.1}$$

where L_t is the total root length at time t, L_i is the amount of infected root per unit soil volume, R is the relative growth rate of the host, dL_i/dt is the rate of formation of new infected root per unit soil volume and $(nL_t - L_i)$ is the amount of infected root per unit soil volume, n being the maximum fraction of the root system which becomes mycorrhizal. S and A are constants. More recently, how-ever, an alternative model has been proposed [9]:

$$\frac{dL_i}{dt} = SL_i\left(1 - \frac{L_i}{nL_t}\right) \tag{6.2}$$

The first model predicts that increased density of both infected and uninfected root, in cm cm^{-3}, will lead to increased encounters be-tween the mycorrhiza and the root. The second model states that only the density of infected root is important until the fraction of the

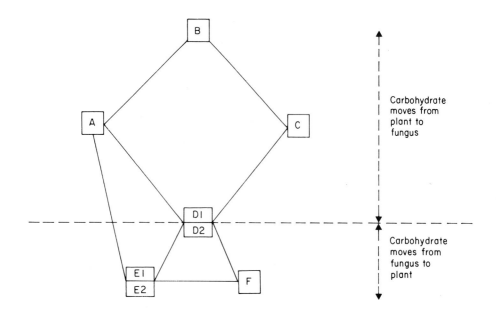

Currently favoured name	Other names
A: Ectomycorrhiza	Ectotrophic, ectocellular, sheathing
B: Vesicular–arbuscular (or Arbuscular)	Endotrophic, phycomycetous
C: Ericoid	Endotrophic ⎫
D1: Arbutoid (with chlorophyllous hosts)	⎪
D2: Monotropoid (with achlorophyllous hosts)	Ectendotrophic ⎬ Ericaceous
E: Orchidaceous	Endotrophic ⎪
E1: With chlorophyllous orchids	⎪
E2: With achlorophyllous orchids	⎭
F: Miscellaneous endo-mycorrhizas*	Endotrophic

*As more information is gathered, mycorrhizal types relegated to F may need to be subdivided or allocated to subgroups within A–E.

All fungi of B, some of F and a few of A are aseptate (? Zygomycetes).
All fungi of C, D and E, most of A and some of F are septate (Ascomycetes or Basidiomycetes).

Fig. 6.7   Interrelationships between various kinds of mycorrhizas. (Data of D. H. Lewis and D. J. Read, Sheffield University.)

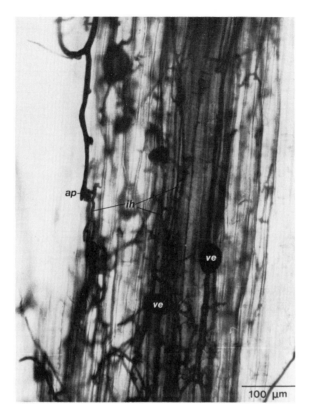

Fig. 6.8   Intercellular hyphae (ih) of yellow vacuolate endophytic mycorrhiza in an onion root, with the entry point of the external hyphae or appressorium (ap) and vesicles (ve). (Photograph by F. E. Sanders, Leeds University.)

root not infected becomes significantly less than 1. Experiments with spring wheat and white clover growing with different numbers of plants per pot in a soil–sand mixture were conducted and the results compared with the output from the two models. Figure 6.9 shows the results for spring wheat grown in the greenhouse at one plant per pot. The second model clearly provides a better fit than the first to the actual results obtained and this was the general case for the various combinations. Such modelling approaches, which are useful for root infections generally, might provide better insight into the endophyte–host interactions.

The relative benefits of infection generally decrease as phosphorus availability increases. The mechanism of the increase in phosphorus uptake is uncertain but the fungi probably explore soil beyond the root surface, solubilize it and transport it to the root. The method of phosphorus transfer to the plant is also uncertain. It has been

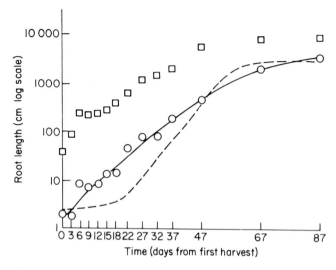

Fig. 6.9   Infected (open circle) and total (open square) root lengths of spring wheat grown in the greenhouse at one plant per pot, with the infected root lengths predicted by Equations 6.1 (broken line) and 6.2 (solid line). (From Buwalda *et al.* [9].)

suggested [36] that the increased concentration of phosphorus and other elements in mycorrhizal plants may be more than that which is apparent from the measurement of carbon/phosphorus ratios in plants, because it depends on their dilution by the photosynthate. Table 6.2 demonstrates that the mycorrhizal plant has 3·3 times more phosphorus than the uninoculated plant, though its dry matter is greater by a factor of only 1·3. It is possible that the mycorrhiza place a drain for carbon from the plant but the significance and magnitude

Table 6.2   Effect of inoculation with ectomycorrhizal fungi on Eastern white pine seedlings. (After Bowen [5].)

	Inoculated	Uninoculated
Dry weight (mg)	405·0	303·0
Root/shoot ratio	0·78	1·04
Nitrogen % dry weight	1·24	0·85
Nitrogen per seedling (mg)	5·00	2·69
Phosphorus % dry weight	0·196	0·074
Phosphorus per seedling (mg)	0·789	0·236
Potassium % dry weight	0·744	0·425
Potassium per seedling (mg)	3·02	1·38
Uptake mg mg^{-1} dry root:		
Nitrogen	0·029	0·016
Phosphorus	0·0045	0·0014
Potassium	0·017	0·008

of this effect has yet to be determined, e.g. if mycorrhizal plants are smaller but take up the same amount of phosphorus as uninfected plants, their phosphorus/carbon ratio will be larger. However, leek plants infected with v–a mycorrhiza attract about 8 per cent more fixed carbon to their roots and appear to have a higher photosynthetic rate than non-infected plants [32]. On a leaf-area basis the net assimilation rate is the same. The infection appears to be effective by increasing the water content of host plants, therefore increasing the leaf area and carbon assimilation and so offsetting the effects of the drain imposed by the mycorrhiza.

The question must also be asked, Do mycorrhizal or nodule symbioses provide something for nothing in terms of energy? It has been indicated [18], for example, that two-thirds of the photosynthate of *Pinus cembra* is diverted to the ectomycorrhiza. Others [29] have also investigated this concept by pulse labelling non-symbiotic and symbiotic faba bean (*Vicia faba*) plants with $^{14}CO_2$ (Fig. 6.10). The mycorrhizal fungus, *Glomus mosseae*, incorporated 1 per cent of the photosynthate and respired 3 per cent. Nodulated plants with *Rhizobium* utilized 7–12 per cent of the photosynthate. However, the legume appeared to compensate in part for the needs of its microbial partners through increased rates of photosynthesis. Unfortunately, these results are a little ambiguous because pulse labelling only gives a partial indication of carbon flow and the specific activity in different

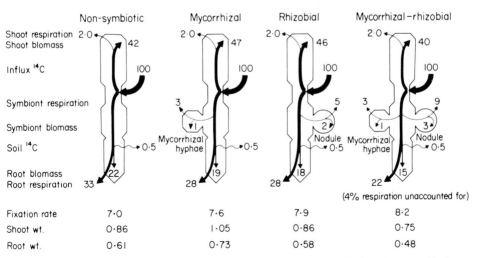

Fig. 6.10   The ^{14}C flow to various compartments of symbiotic and non-symbiotic faba beans (four to five weeks old) after shoots were exposed above ground to $^{14}CO_2$ under continuous light. The shoot weight and the root weight are expressed as grams of carbon. The carbon flux has been equalized to 100 units of carbon per gram of shoot carbon. (From Paul & Kucey [29].)

parts of the plant may vary. It is better to grow plants exclusively on a source of $^{14}CO_2$. The ^{14}C approach is very useful and should answer the important question of energy transfer.

6.4   **Plant pathogens**

The majority of plant pathologists are concerned with foliar pathogens which cause disease symptoms; these are usually relatively easy to recognize compared with root diseases. Many of these foliar diseases, such as eyespot in wheat, are soil borne but it would be totally beyond the scope of this book to consider any of them in detail. It is enough to emphasize that when the pathogen is soil borne, the ecological principles outlined in earlier chapters for microorganisms are generally also relevant to the study of pathogens.

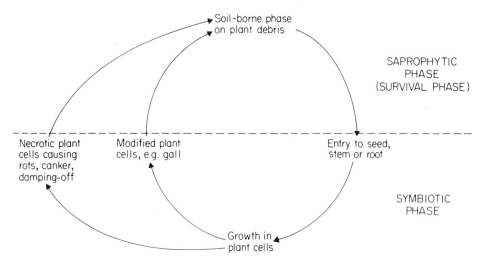

Fig. 6.11 Life cycle of a typical root pathogen.

Fig. 6.12 Some root diseases. (a) Club root (*Plasmodiophora brassicae*) of cabbage. Initial wilting gives rise to stunting and sometimes plants die completely; survivors do not thrive. The roots show swellings and contortions which have a mottled appearance when sectioned and ultimately they rot. (Photograph by ICI Plant Protection Division.) (b) Clover rot (*Sclerotinia trifoliarum*). This is one cause of clover sickness when red clover fails due to being grown too frequently. Stems and foliage are attacked first, followed by root rot with subsequent plant death. (Crown Copyright.) (c) Violet root-rot (*Helicobasidium purpureum*) of sugar-beet. The roots become infected with this violet or purple-brown fungus but there can also be secondary bacterial decay. The mycelium contains raised spots or pin-heads, known as 'infection cushions', and sclerotia. (Crown Copyright.) (d) Crown gall (*Agrobacterium radiobacter* var. *tumefaciens*) of sugar-beet. This bacterial disease causes characteristic swellings. It also affects almond, peach and rose where it can be controlled biologically (section 10.4). (Crown Copyright.)

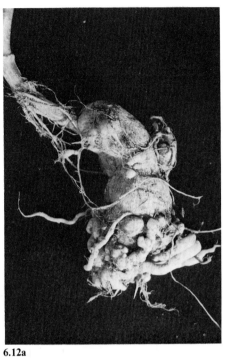

6.12a

6.12b

6.12c

6.12d

It is, however, important that the microbiologist recognizes that pathogens usually have distinct life cycles, often with saprophytic and parasitic phases (Fig. 6.11). A summary of root diseases can be found in the book by Krupa and Dommergues [24].

Some diseases produce distinct anatomical changes in the host (Fig. 6.12) but some of the cereal-root diseases (Fig. 6.13) can be difficult to distinguish, e.g. in the early stages of infection with take-all, there is merely a browning of the root system which can easily be confused with *Fusarium* root-rots or other disorders. In the later stages, the symptoms become quite distinctive and only take-all produces 'whiteheads' in the grain. Root invasion by take-all is shown in Fig. 6.14. The effects of root-infecting fungi on the structure and function of cereal roots have been discussed [21] but some of the effects of pathogens can be more subtle than is recognized by normal field observation, e.g. it has been shown [22, 23] in solution culture

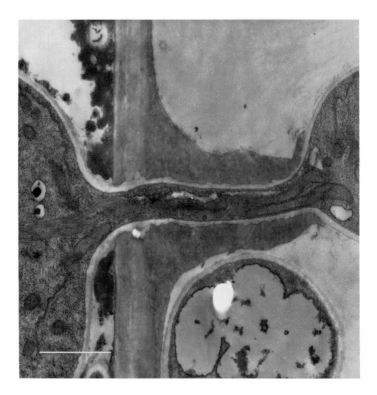

Fig. 6.14   Penetration of a cortical cell wall of wheat by the take-all fungus. The cytoplasm is necrotic and a lignotuber has been formed at the point of entry. The hole in the cell wall has been enlarged beyond the size needed for the passage of the hyphae, suggesting enzymic action. Bar marker = 1 μm. (Photograph by J. L. Faull & R. Campbell, Bristol University.)

(a)

(b)

(c)

(d)

Fig. 6.13 Some soil-borne diseases of wheat.
(a) Eyespot (*Pseudocercosporella herpotrichoides*).
This can also occur on barley but oats are
resistant. Note the pale, oval spot with a brown
margin; this can occur on basal leaf sheaths and
culm of cereal tillers or just above the ground.
Unlike sharp eyespot (*Rhizoctonia solani*), the
elliptical, eye-shaped lesions have a diffuse margin
and a dark pupil with grey mycelium in the
internal cavity of the straw. (Copyright by ICI
Plant Protection Division.) (b) Foot-rot (*Fusarium
culmorum*). This fungus also causes seedling blight
and is difficult to distinguish on the plant from *F.
avanaceum* and *F. nivale*. It therefore has to be

isolated. (Copyright by ICI Plant Protection
Division.) (c) Take-all or 'whiteheads'
(*Gaeumannomyces graminis*). There are two
varieties—*tritici* and *avenae*. Seedlings are often
killed and survivors are stunted with fewer tillers.
Emerged heads have empty spikelets and appear
blanched (whiteheads). The reduced and blackened
roots are easily pulled from the ground.
(Photograph by R. D. Prew, Rothamsted
Experimental Station.) (d) Root-rot (*Fusarium
culmorum*). This is similar to foot-rot but lesions
form on the roots instead of the stem base.
(Photograph by R. D. Prew, Rothamsted
Experimental Station.)

studies that two unidentified organic acids from *Fusarium culmorum* inhibited root-hair formation in barley (Fig. 6.15) and retarded growth but increased the stomatal diffusion resistance, chlorophyll content of shoots and phosphorus uptake. The rate of photosynthesis per unit area of leaf and the total carbon and nitrogen content of plants did not change. Both *Fusarium* spp. and *Pythium* spp. contribute to the phenomenon of 'damping off', which can be difficult to diagnose precisely because plants either fail to emerge or 'die back' shortly after emergence. This is quite common, especially with sugar-beet, lettuce and grasses.

Pythium infections can also result in further symptoms in cereals which the field pathologist finds difficult to recognize. All stages of plant development, including tillering, can be inhibited. The stems are of uneven height producing a plant which generally looks weak (Fig. 6.16). Figure 6.17 shows oospores of a *Pythium* sp. in a fine rootlet of wheat; the invasion causes the roots to cease growth and function. The disease has been identified indirectly by eliminating it after fumigating soil with methyl bromide or applying N-(2,6-dimethylphenyl)-N-(methoxyacetyl)-alanine methyl ester, a compound specific against *Pythium* and related fungi [10]. *Pythium ultimum* was recovered from more than 50 per cent of the plants in non-treated plots.

The root pathologist recognizes some important concepts of which the soil microbiologist should be aware [15,40]:

1. *Koch's postulates.* These are criteria which were established by Henle in 1840 to establish the causal relationship between a microorganism and a disease and were first satisfied by Koch in 1876 with his experiments on anthrax. They are: (a) the micro-organism must be present in every case of the disease; (b) the micro-organism must be isolated from the diseased host and grown in pure culture; (c) the specific disease must be reproduced when the culture is inoculated into a healthy, susceptible host; and (d) the micro-organism must be recoverable from the experimentally infected host. In practice, it is common that not all these conditions are satisfied because there can be complicating factors, e.g. the organism grown in culture may lose its pathogenicity.

2. *Specialization of the pathogen.* There are three categories. The first is that of the unspecialized, root-infecting fungi. Host resistance normally limits infection to juvenile tissues but some tissues are predisposed to infection by some environmental factor adverse to the general health of the host crop. Phytophthora infections of trees are good examples. The second category is that of the vascular-wilt fungi which, although their entry is restricted to juvenile tissues, are highly specialized in their infection habit. The infection route is

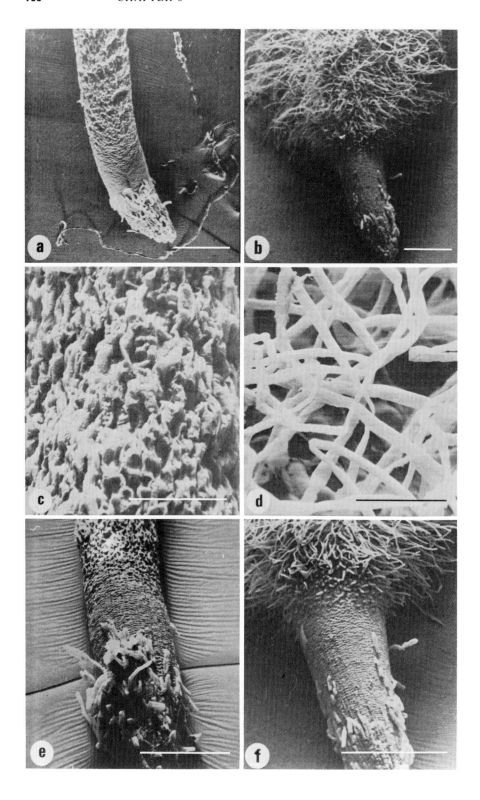

Fig. 6.16 Lettuce plant 30 days after inoculation with *Pythium* isolate L4. The plant on the left shows severe stunting and the plant on the right is the uninoculated control (Coplin *et al.* [11]). (Photograph by D. L. Coplin, Ohio Agricultural Research and Development Center.)

usually via the immature cortex of the apical region of the rootlets and endodermis, into the xylem tracts and then systemically through the entire vascular system of the plant. Thus, the number of primary infection sites is less important than in the unspecialized fungi. *Fusarium* and *Verticillium* infections are good examples. The third is that of the ectotrophic root-infecting fungi. Advance of the fungus is, at first, epiphytic over the surface of the host and then infection hyphae, either singly or in aggregates, invade the underlying host tissues. Their specialization lies between the two categories above and examples are take-all and the root- and butt-rot of conifers caused by *Fomes annosus*. The number of primary infections and favourable conditions for the infections at the soil–root interface is

Fig. 6.15 Scanning electron micrographs of roots of barley seedlings grown in Hoagland nutrient medium with (a, c and e) and without (b, d and f) the addition of the acids produced by *Fusarium culmorum*. Bar markers = 1 mm. (Photographs by M. Katouli and R. Marchant, New University of Ulster.) (a) Low magnification of a root, treated with 450 g ml^{-1} of the organic acids, in nutrient medium adjusted to pH 6·2. Note the absence of root hairs. (b) Low magnification of a root of a normal barley seedling grown in Hoagland medium (pH 6·2). Note the abundance of long root hairs. (c) Higher magnification of a treated barley root. Note that older epidermal cells have initiated some root hairs. (d) Root hairs of a normal root grown in Hoagland medium at a higher magnification. Notice the difference from (c) in the same region of the root. (e) Elongation zone and the root cap of a treated barley seedling. (f) Root cap and elongation-zone cells of a root of a normal barley seedling.

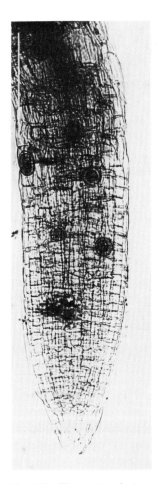

Fig. 6.17   Fine rootlet of wheat containing oospores of a *Pythium* sp. Bar marker = 135 μm. (Photograph by J. Sitton and C. Chamswarng, Washington State University.)

critical in establishing a progressive ontogeny of disease, decreases with increasing specialization of the pathogen.

3. *Infection court.*   This region is the initial site of contact between the pathogen and the surface of the host.

4. *Inoculum potential.*   This is the 'fungal energy for growth' and is related to the volume and nutrient status of the inoculum and to environmental factors at the soil–root interface. The term is merely conceptual and therefore rather 'loose' in comparison with the colonization potential as applied to the rhizosphere (section 5.3).

5. *Biological balance.*   In analysing plant disease, many philosophical concepts, some of which do not lend themselves to quantitative analysis, need to be considered. As an illustration, the following

principles contributing to the biological balance of soil are emphasized by Baker and Cook [2]:

a. An organism will increase until the limitations imposed by the biotic and abiotic environment just counterbalance the increase.

b. The more favourable the environment to the pathogen and the more severe the selection pressure on the host, the higher the level of resistance maintained in the host.

c. In the absence of the host or adequate nutrients, a pathogen is apt to survive longest in soil under non-lethal conditions favourable to germination.

d. Parasites are at a greater competitive disadvantage than saprophytes when they are outside the host.

e. Possession is nine points of the law for micro-organisms in relation to host tissue.

f. Environmental conditions which are relatively more favourable to facultative types of pathogens than to the host, may or may not favour increased infection but generally increase damage. Conditions most favourable for the host generally are most favourable for infection and damage by obligate parasites.

g. The progression through the stages of the life cycle of a soil micro-organism is determined at least as much by the associated microflora and abiotic environment as by the genes of the micro-organisms.

h. The greater the complexity of the biological community, the greater its stability.

Whereas all the ideas are useful as working hypotheses, they really need to be tested experimentally because there are often good counter-arguments, e.g. some biological communities, such as those in the rumen, have few members and yet are stable. Similar small communities resistant to external influences have been mimicked in the chemostat.

6. *Factors influencing disease.*   All the soil chemical and physical factors (Chapter 2) influence infection and ontogeny of disease. Of well proven importance are the osmotic and matric potentials— temperature and aeration, e.g. with normal aeration of soil, *Fusarium solani* f.sp. *pisi* had a negligible effect on bean growth, whereas *F. solani* f.sp. *phaseoli* severely injured the plants [27]. After a temporary imposition of low soil-oxygen levels, the pea pathogen (*F. solani* f.sp. *pisi*) also injured the bean roots and permanently reduced water absorption and plant growth, although the injury was less than that caused by the bean pathogen. This shows that the temporary oxygen stress predisposes the host to infection by that particular pathogen.

Chemotaxis (sections 5.1 and 6.2) to the roots may be a trigger for disease initiation.

7. *Phytoalexins.* When an avirulant race infects plant tissue, persistent substances—phytoalexins—can be produced by the plant and prevent infection by virulent races. This is a hypersensitive response by the plant. Phytoalexins are not particularly selective in their action, usually being as toxic to non-pathogens as pathogens. Some other forms of plant damage, including that caused by chemicals, can induce phytoalexin production. Whereas the evidence for production of these substances in foliar tissues is quite good, that for production in roots is less clear. As they are produced in such small quantities and large amounts of non-sterile roots have to be extracted to identify them, it is possible that they could originate from the microflora associated with the root.

8. *Suppressive soils.* Some soils are not conducive to particular diseases, while in others disease declines after a period. Take-all decline is when the disease peaks in the third or fourth year of continuous wheat crops and then declines. Sometimes soil can be transferred to make conducive soil become suppressive. This is thought to be due to the transfer of microbial antagonists of the pathogen (section 10.3). Soil mineral composition (Chapter 2) can influence these effects, e.g. montmorillonite clay in soil promotes the growth of bacteria which suppresses *Fusarium oxysporum*, the cause of Panama disease of banana.

9. *Iatrogenic plant diseases.* Diseases which result from, or are increased by, the use of a specific crop-protection chemical may be referred to as iatrogenic, a term derived from human medicine [17].

6.5 **Conclusion**

Symbiotes and pathogens are soil micro-organisms with habits of proven economic significance in terms of plant productivity. As such, their genetics, physiology and ecology have been studied in greater detail than any other group of soil micro-organisms. However, the explanation of the 'soil sickness' phenomenon of clover, when a legume fails to nodulate or does so inefficiently, is still somewhat unclear. The beneficial and adverse infections of plants need to be considered in relation to microbial activities which do not result from infections.

References

1 Agrios G. N. (1973) *Plant Pathology.* Academic Press, New York.
2 Baker K. F. & Cook R. J. (1974) *Biological Control of Plant Pathogens.* W. H. Freeman, San Francisco.
3 Beyon J. L. & Josey D. P. (1980) Demonstration of heterogeneity in a natural population of *Rhizobium phaseoli* using variation in intrinsic antibiotic resistance. *Journal of General Microbiology*, **118**, 437–42.
4 Bezdicek D. F. & Donaldson M. D. (1980) Flocculation of *Rhizobium* from soil

colloids for enumeration by immunofluorescence. In *Microbial Adhesion to Sur-faces*, eds Berkely R. C. W., Lynch J. M., Melling J., Rutter P. R. & Vincent B., pp. 297–309. Ellis Horwood, Chichester.

5 Bowen G. D. (1973) Mineral nutrition of ectomycorrhizae. In *Ectomycorrhizae: their Ecology and Physiology*, eds Marks G. C. & Kozlowski T. T., pp. 151–205. Academic Press, New York.

6 Burns R. C. & Hardy R. W. F. (1975) *Nitrogen Fixation in Bacteria and Higher Plants*. Springer-Verlag, New York.

7 Brill W. J. (1981) Agricultural microbiology. *Scientific American*, **245**, 199–215.

8 Broughton W. J. (1981) *Nitrogen Fixation. Vol. 1 Ecology*. Clarendon Press, Oxford.

9 Buwalda J. G., Ross G. J. S., Stribley D. P. & Tinker P. B. (1982) The development of endomycorrhizal root systems. III The mathematical representation of the spread of vesicular–arbuscular mycorrhizal infection in root systems. *New Phytologist*, **91**, 669–82.

10 Cook R. J., Sitton J. W. & Waldher J. T. (1980) Evidence for *Pythium* as a pathogen of direct-drilled wheat in the Pacific Northwest. *Plant Disease*, **64**, 102–3.

11 Coplin D. L., Schmitthenner A. F. & Bauerle W. L. (1980) Root rot of lettuce incited by *Pythium polymastum*. *Plant Disease*, **64**, 63–6.

12 Dazzo F. B. & Hubbell D. H. (1975) Cross-reactive antigens and lectin as deter-minants of symbiotic specificity in the *Rhizobium*–clover association. *Applied Microbiology*, **30**, 1017–33.

13 Evans H. J., Emerich D. W., Maier R. J., Hanus F. J. & Russell S. A. (1979) Hydrogen cycling within the nodules of legumes and non-legumes and its role in nitrogen fixation. In *Symbiotic Nitrogen Fixation in the Management of Temperate Forests*, eds Gordon J. C. J., Wheeler C. T. & Perry D. A., pp. 196–206. Oregon State University Press, Corvallis.

14 Garrett S. D. (1970) *Pathogenic Root-infecting Fungi*. Cambridge University Press, Cambridge.

15 Garrett S. D. (1979) The soil–root interface in relation to disease. In *The Soil–Root Interface*, eds Harley J. L. & Russell R. S., pp. 301–3. Academic Press, London.

16 Gaworzewska E. T. & Carlile M. J. (1982) Positive chemotaxis of *Rhizobium leguminosarum* and other bacteria towards root exudates of legumes and other plants. *Journal of General Microbiology*, **128**, 1179–88.

17 Griffiths E. (1981) Iatrogenic plant diseases. *Annual Review of Phytopathology*, **19**, 69–82.

18 Harley J. L. (1969) *The Biology of Mycorrhiza*, 2nd edn. Leonard Hill, London.

19 Hayman D. S. (1978) Endomycorrhizae. In *Interactions Between Non-pathogenic Soil Micro-organisms and Plants*, eds Dommergues Y. R. & Krupa S. V., pp. 401–42. Elsevier, Amsterdam.

20 Hirrel M. C., Mehravaran H. & Gerdemann J. W. (1978) Vesicular–arbuscular mycorrhizae in the Chenopodiaceae and Cruciferae: do they occur? *Canadian Journal of Botany*, **56**, 2813–17.

21 Hornby D. & Fitt B. D. L. (1981) Effects of root-infecting fungi on structure and function of cereal roots. In *Effects of Disease on the Physiology of the Growing Plant*, ed. Ayres P. G., pp. 101–30. Cambridge University Press, Cambridge.

22 Katouli M. & Marchant R. (1981) Effect of phytotoxic metabolites of *Fusarium culmorum* on the growth and physiology of barley plants. *Plant and Soil*, **60**, 377–84.

23 Katouli M. & Marchant R. (1981) Effect of phytotoxic metabolites of *Fusarium culmorum* on barley root and root-hair development. *Plant and Soil*, **60**, 385–97.

24 Krupa S. V. & Dommergues Y. R. (eds) (1979) *Ecology of Root Pathogens.* Elsevier, Amsterdam.

25 Lewis D. H. (1974) Micro-organisms and plants: the evolution of parasitism and mutualism. In *Evolution in the Microbial World*, eds Carlile M. J. & Skehel J. J., pp. 367–92. Cambridge University Press, Cambridge.

26 Marx D. H. & Krupa S. V. (1978) Ectomycorrhizae. In *Interactions Between Non-pathogenic Soil Micro-organisms and Plants*, eds Dommergues Y. R. & Krupa S. V., pp. 373–400. Elsevier, Amsterdam.

27 Miller D. E., Burke D. W. & Kraft J. M. (1980) Predisposition of bean roots to attack by the pea pathogen, *Fusarium solani* f.sp. *pisi*, due to temporary oxygen stress. *Phytopathology*, **70**, 1221–4.

28 Nutman P. S. (1976) *Symbiotic Nitrogen Fixation in Plants.* Cambridge University Press, Cambridge.

29 Paul E. A. & Kucey R. M. N. (1981) Carbon flow in plant microbial associations. *Science*, **213**, 473–4.

30 Powell C. (1976) Mycorrhizal fungi stimulate clover growth in New Zealand hill country soils. *Nature (London)*, **264**, 436–8.

31 Schippers B. & Gams W. (1979) *Soil-borne Plant Pathogens.* Academic Press, London.

32 Snellgrove R. C., Splittstoesser W. E., Stribley D. P. & Tinker P. B. (1982) The distribution of carbon and the demand of the fungal symbiont in leek plants with vesicular–arbuscular mycorrhizas. *New Phytologist*, **92**, 75–87.

33 Sprent J. I. (1979) *The Biology of Nitrogen-fixing Organisms.* McGraw-Hill, Maidenhead.

34 Starr M. P. (1975) A generalized scheme for classifying organismic association. In *Symbiosis*, eds Jennings D. H. & Lee D. L., pp. 1–20. Cambridge University Press, Cambridge.

35 Stewart W. D. P. (ed.) (1975) *Nitrogen Fixation by Free-Living Organisms.* Cambridge University Press, Cambridge.

36 Stribley D. P., Tinker P. B. & Rayner J. H. (1980) Relation of internal phosphorus concentration and plant weight in plants infected with vesicular–arbuscular mycorrhizas. *New Phytologist*, **86**, 261–6.

37 Tinker P. B. H. (1975) Effects of vesicular–arbuscular mycorrhizas on higher plants. In *Symbiosis*, eds Jennings D. H. & Lee D. L., pp. 325–49. Cambridge University Press, Cambridge.

38 Zablotowicz R. M. & Focht D. D. (1979) Denitrification and anaerobic, nitrate-dependent acetylene reduction in cowpea *Rhizobium*. *Journal of General Microbiology*, **111**, 445–8.

39 Zamir A., Maina C. V., Fink G. R. & Szalay A. A. (1981) Stable chromosomal integration of the entire nitrogen-fixation gene cluster from *Klebsiella pneumoniae* in yeast. *Proceedings of the National Academy of Sciences of the USA*, **6**, 3496–500.

40 Zentmyer G. A. (1979) Effect of physical factors, host resistance and fungicides on root infection at the soil-root interface. In *The Soil–Root Interface*, eds Harley J. L. & Russell R. S., pp. 315–28. Academic Press, London.

PLANT GROWTH REGULATORS AND PHYTO-TOXINS FROM MICRO-ORGANISMS

Often again it profits to burn the barren fields, firing their light stubble with crackling flame: It is whether the earth conceives a mysterious strength and sustenance thereby, or whether the fire burns out her bad humours and sweats away the unwanted moisture or whether the heat opens more of the ducts and hidden pores by which her juices are conveyed to the fresh vegetation – or rather hardens and binds her gaping veins against fine rain and consuming sun's fierce potency and the piercing cold of the north wind.

Virgil   *The Georgics* 1, 84–93, 37 B.C. (Translation by C. Day Lewis.)

We have seen in the preceding chapter that it is difficult to classify a micro-organism as a pathogen. Some micro-organisms causing distinct disease symptoms are recognized as pathogens. Others can inhibit plant growth, particularly by producing phytotoxic metabolites, yet are not generally considered to be pathogens. The latter group might be termed 'subclinical pathogens', 'minor pathogens' or 'deleterious organisms' and will form the second major focus of this chapter. Micro-organisms which compete with seeds and roots for available oxygen (section 2.3) might also be considered in this group.

The study of allelopathy [23] concerns the inhibition of the growth of one plant species by another, including the indirect involvement of rhizosphere micro-organisms from the inhibitor species or saprophytes producing toxins from plant residues. Most of the substances referred to in this chapter can therefore be termed allelochemicals.

7.1   Products of soil micro-organisms

Micro-organisms can produce a vast range of metabolites, probably several tens of thousands [12]. Therefore, to identify any one metabolite as being an active phytotoxin might be compared with searching for the proverbial 'needle in a haystack'. However, this search need not be so complex if it is recognized that for the product to be of ecological significance it must be biologically active in the form and concentration in which it is actually present. Relatively few investigators have satisfied these conditions. Many compounds, particularly polymers, can be degraded if acid or alkali are used as the extractants from soils, micro-organisms or plants. Extraction of metabolites in water avoids this complication.

The concentration can be difficult to assess because metabolites are usually only produced at the site of microbial activity, yet often

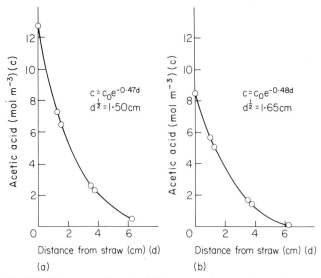

Fig. 7.1 Movement of acetic acid from straw decomposing under anaerobic conditions. (a) Barley straw; (b) wheat straw. $d^{\frac{1}{2}}$ is the distance through which c is reduced by a factor of 2. (After Lynch *et al.* [15].)

the bulk soil is extracted. Allowance for the 'dilution factor' must be made, e.g. acetic acid is produced from straw by microbial fermentation but it does not diffuse very far (Fig. 7.1). Plant damage by this phytotoxic substance occurs only when the seedling roots touch or come very close to the decomposing straw. Therefore, the site, as well

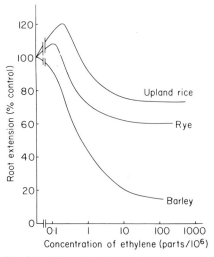

Fig. 7.2 Effect of ethylene on the root extension of cereals. (After Smith & Robertson [27].)

as the intensity of the activity, must be considered. A further aspect of the concentration effect is that compounds are commonly inhibitory in one concentration while at lower concentrations they are stimulatory (Fig. 7.2). Most of the plant growth regulators behave in this way.

7.2   **Plant growth regulators**

Examples of plant growth regulating chemicals are shown in Fig. 7.3.

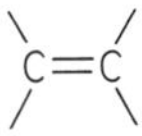

Ethylene

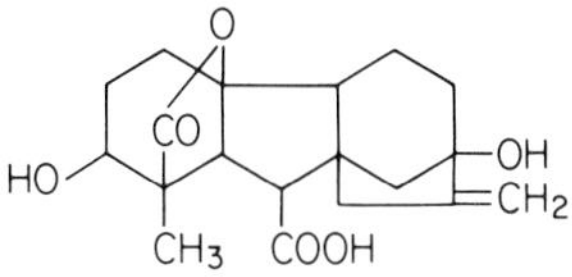

Gibberellic acid (GA₃)

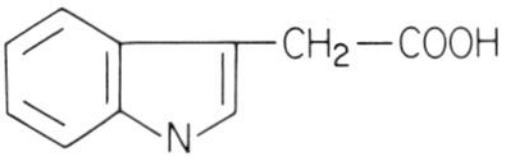

Auxin (IAA, indolyl-3-acetic acid)

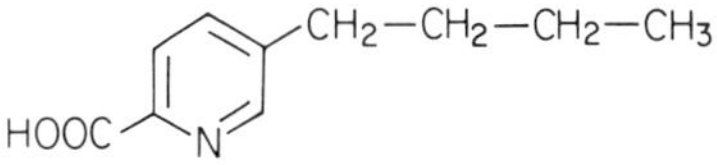

Fusaric acid
(5-n-butyl picolinic acid)

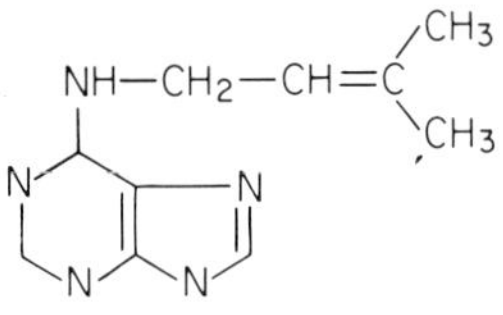

Isopentyladenine
(6-(γ,γ-dimethylallyl)-aminopurine)

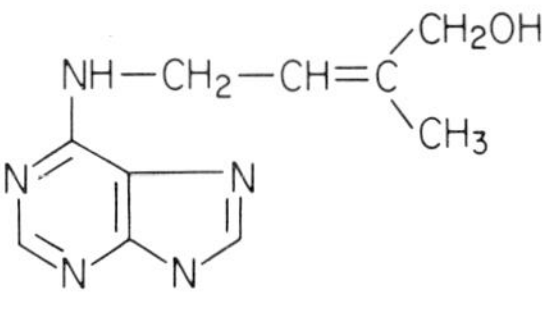

Zeatin
(6-(4-hydroxy-3-methylbut-2-enyl) aminopurine)

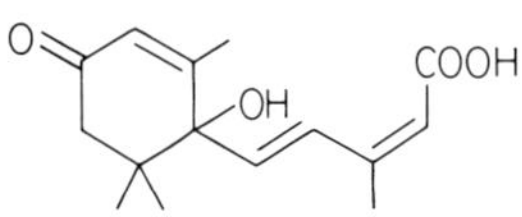

Abscisic acid (ABA)

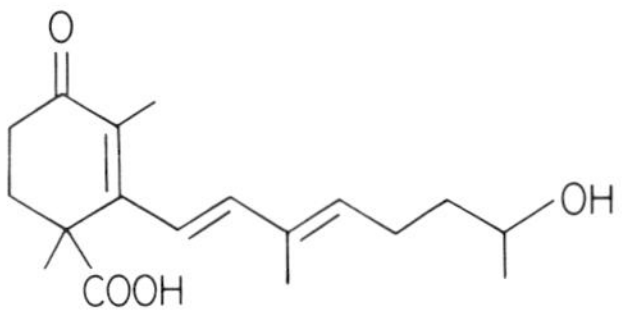

Trisporic acid 'C'

Fig. 7.3 Plant growth regulators.

7.2.1.   *Ethylene*

This gas was first identified as a plant growth regulator in 1901. In the past decade there has been a great interest in its production and effects. Its many effects on plant growth are well documented [1]. Interest to soil microbiologists stems from the observations [28] that in laboratory experiments using anaerobic conditions, ethylene could occur in the soil atmosphere at concentrations sufficient to inhibit extension growth of cereal roots (Fig. 7.4). The gas seemed to be the

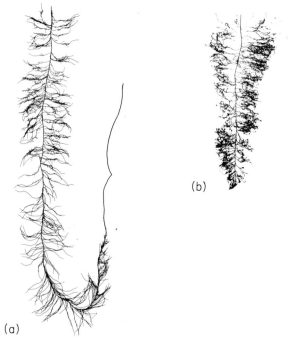

(b)

(a)

Fig. 7.4   Effects of ethylene on the growth of barley roots. (a) Control; (b) roots exposed to 100 parts/10^6 ethylene for 25 days.

result of microbial activity because it was not produced after autoclaving [26]. It was shown that methionine was the substrate for the formation of the gas and that saprophytic fungi were the producers [11]. However, there appeared to be a paradox—low oxygen concentrations favoured the formation of the gas in soil [26] and oxygen favoured the process in pure cultures of *Mucor hiemalis* [16]. This paradox was subsequently resolved by the observations that aerobic conditions favour the process in soil when substrates, originating from crop residues, are in excess and that the breakdown of the gas to ethylene oxide is also favoured aerobically, thus restricting the accumulation of the gas [17]. A biochemical intermediate, possibly the keto-acid, accumulates extracellularly and this can break down abiotically to ethylene, the latter process being catalysed by Fe^{2+}. A similar process was subsequently described for bacteria [22] but as their biomass in soil is smaller than that of fungi, it seems likely that primary saprophytic sugar fungi are better competitiors for substrates.

These observations were disputed [25] because it was claimed that ethylene is fungistatic and generally microbiostatic to aerobes. The concept of fungistasis is applied when the germination and growth of fungi in natural soils are much more restricted than would be expected from their behaviour under similar conditions of temperature, moisture and pH *in vitro*. The evidence for this was recently extensively reviewed [10]. Similar observations were subsequently made for bacteria. Demonstration of fungistasis has usually been by measuring percentage spore germination after the incubation of agar discs placed in soil and comparing it with that on filter paper. Generally, sterilization of the soil has alleviated the inhibition. Two possible explanations, which are not necessarily mutually exclusive, are the presence of toxins in the soil or an inadequate nutrient supply.

In the oxygen–ethylene cycle [25], organic matter in soil allowed aerobic growth, which depleted the oxygen supply, resulting in anaerobic conditions. Ferrous iron, which is the trigger for ethylene production, was produced anaerobically. The ethylene diffused to inhibit the growth of aerobes. This resulted in less oxygen demand, oxygen diffusing back to the system, inhibiting anaerobes, aerobes growing and the cycle repeating itself. The major weakness in this otherwise attractive argument, is that the gas does not appear to be generally fungistatic and ethylene production by an anerobe has never been detected. Present evidence indicates that the fungal origin is still most likely in soil, although in the rhizosphere aerobic bacteria are probably important. Furthermore, whereas ammonia is clearly a fungistatic volatile in alkaline soils [9], the overriding evidence now suggests that nutrient deprivation is the major factor in microbiostasis [14] and that the inhibition of microbial growth in soil can be explained in simple terms of growth theory (Chapter 4) because inadequate substrates are available.

7.2.2   *Gibberellins*

In 1926, Kurosawa in Taiwan noted that a fungus associated with rice plants caused the plants to die but that prior to this they became abnormally tall. The fungus is now known to be *Gibberella fujikuroi*, the perfect (i.e. producing sexual spores) state of *Fusarium maniliforme*, and the active principle to be gibberellic acid, an isoprenoid synthesized from mevalonic acid. Now over 50 gibberellins are known, all with the same basic chemical structure (gibbanes). Most of the gibberellins have been found in plants and it is perhaps surprising that few have been found in micro-organisms. This may be because they have not been sought. Indeed, few agrochemical companies have heeded the message from Kurosawa's observation and instead of hunting for plant growth regulators in micro-organisms, they have commissioned the chemist to synthesize them. This is in marked contrast to the pharmaceutical industry which has searched for antibiotics from soil micro-organisms extensively.

It would appear that gibberellin-like compounds, as determined by bioassays, can be produced by rhizosphere bacteria [4], but rigorous chemical determination and dose-response evidence is lacking. Certainly, the bioassay data could be misleading as even *G. fujikuroi* produces another inhibitor of plant growth—fusaric acid (5-n-butyl picolinic acid). The balance between growth promotion and inhibition could be delicate and this could be a cause of the odd effects when *Fusarium* is inoculated onto plants (Chapter 6).

7.2.3   *Auxin*

Indol-3yl-acetic acid (IAA) is a biosynthetic product from tryptophan and it is one of the most common examples of the group of substances often structurally unrelated but known generally as auxins. As the enzymic step from tryptophan is relatively facile and the enzyme is widely distributed, it is not surprising that many micro-organisms can carry out this reaction. However, the problem then becomes one of trytophan availability in soil and around roots and whether sufficient can be converted to cause an effect on the growth of the plant. There have been no convincing demonstrations of this. Auxin inhibits the elongation of roots with a concurrent thickening. Lateral growth, which is stimulated, does not compensate for the dry matter lost. Low concentrations or brief exposure to auxin can be stimulatory to root growth. Many effects of auxin are caused indirectly by a stimulation of ethylene formation in the plant. However, most of these effects are of endogenous IAA and there have been only a few reports of studies with exogenously applied IAA.

Hammence [8] found around 5 ng IAA $100 g^{-1}$ soil but others [30] could detect only phenolic acids with auxin-like activity in soil. IAA can be readily degraded by micro-organisms in soil and therefore it probably has greater significance to the plant when formed symbiotically by mycorrhiza or nodules.

7.2.4   *Cytokinins*

Cytokinins are aminopurines and as such are linked to phosphate esters of sugars to form nucleosides and nucleotides. They promote cell division in plant tissues. Zeatin appears to be the most common cytokinin of plants and isopentyladenine the most common of micro-organisms [29]. As yet, little evidence has been produced to indicate that cytokinin production by micro-organisms has any significance.

7.2.5   *Abscisic acid*

Abscisic acid (ABA) is primarily a plant growth inhibitor. It is also involved in abscission, dormancy of seeds and ion uptake by roots. Biosynthetically, it probably originates from mevalonate but it could also arise from fragmentation of carotenoids. It has not yet been reported as a microbial product but this probably merely reflects that it has not been looked for. Mucoraceous fungi can produce trisporic acids, which have superficial structural similarities to ABA when (+) and (−) sexual forms mate [2]. Their effects on roots is somewhat similar to ABA [18] but because the carbon double bonds are fully conjugated in the trisporic acids, the compounds are unstable and activity may be due to breakdown products.

7.3   **Phytotoxic substances**

It is sometimes difficult to distinguish phytotoxins from plant growth regulators. Phytotoxins are less commonly transported within the plant and are sometimes only active in greater concentrations than growth regulators. Pathologists usually reserve the term phytotoxin for secondary metabolites which accumulate in low concentrations, on the basis that any chemical can be toxic when present in large concentrations. Soil scientists, however, generally apply the term to any metabolite, primary or secondary, which accumulates in toxic concentrations in soil and this is the usage here. The occurrence of phytotoxins in soil probably originates from microbial degradation processes but plant enzymes might also be responsible to some extent. Certainly, micro-organisms have the capacity to produce most of the identified phytotoxins but there has seldom been a rigorous distinction between the two potential sources in studies on their formation.

7.3.1   *Organic acids*

One of the earliest investigations of phytotoxins in soil was by Schreiner and Skinner [24] who attempted to isolate them from non-fertile soils. Using classical chemical techniques they identified dihydroxystearic acid:

$$CH_3(CH_2)_7.CHOH.CHOH(CH_2)_7 COOH.$$

The acid was most active when the plant was under nutrient stress but the boundary conditions outlined in section 7.1 were not satisfied. Other work in the same laboratory showed phytotoxic effects of α-crotonic acid, salicyladehyde, picolinic acid and vanillin. Subsequent studies by others also considered tyrosine, salicylic acid, coumarin, 5-hydroxy-naphthoquinone (juglone) and cinnamic acid as phytotoxins, but without convincing evidence.

In more recent years, aliphatic acids which are volatile in steam (acetic, butyric and propionic) have been demonstrated as likely phytotoxins in soils, particularly when the soil becomes anaerobic. One of the first convincing studies was in Philippine soils [5], although the acids were extracted with phosphonic acid and therefore may not have been free. Later workers using water extracts found them in the free state in soils, particularly when straw or green manure had been added to the soil. This is not surprising, the acids probably forming during the microbial fermentation of cellulose (section 9.3).

The effect on seedling root growth of the organic acids when applied at seeding to barley grown on sand is shown in Fig. 7.5. The

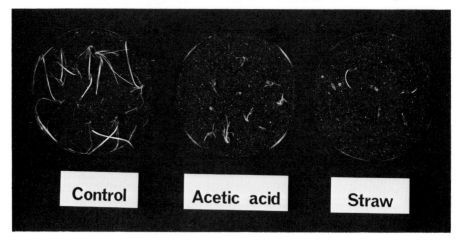

Fig. 7.5   Effect of acetic acid ($15 \, mol \, m^{-3}$) and a solution from fermented wheat straw containing a similar concentration of acetic acid on the early seedling growth of barley.

effect varies between different plants but, generally, phytotoxicity occurs at concentrations similar to those which affect barley, although rice seems more sensitive. The aliphatic acids are most phytotoxic at low pH (Table 7.1) because then they are lipophilic and soluble in the lipid components of the root membranes [13]. This is because the organic acids are weak acids and dissociate as follows:

$$CH_3COOH \rightleftharpoons CH_3COO^- + H^+.$$

The pKa of acetic acid, for example, is 4·75 and therefore 86 per cent of the acid is undissociated at pH 4, whereas at pH 6 only 5 per cent is undissociated. Thus, under alkaline conditions the phytotoxicity is minimal and this has opened an avenue for the control of their activity (section 10.1).

Aromatic acids (Fig. 7.6) can also be formed in soils and, although they are at least as toxic as the aliphatics (Table 7.1), there has never been a clear demonstration that they can reach phytotoxic

Table 7.1   Effect of some organic acids at a concentration of $5\,mol\,m^{-3}$ on the root extension of barley seedlings*. (From Lynch [13].)

Acid	Root extension (% control)†	
	pH 6·5	pH 5·5
Aliphatic		
Acetic	100	75
Butyric	34	29
Citric	105	74
Formic	61	80
Lactic	94	103
Propionic	48	50
Succinic	73	81
Aromatic		
Benzoic	69	26
p-coumaric	25	17
p-hydroxybenzoic	7	57
3-phenylpropionic	27	25
Salicylic (o-hydroxybenzoic)	55	26
Syringic (3, 5-dimethoxy-4-hydroxybenzoic)	55	53
Vanillic (3-methoxy-4-hydroxybenzoic)	62	58
Amino		
Glycine	101	92
DL-methionine	62	90
DL-tyrosine	85	95

* The percentage of germination was between 93 and 100% in all treatments except benzoic and salicylic acids at pH 5·5, where it was 40%.
† The least significant difference at $P = 0.05$ was 15% of control.

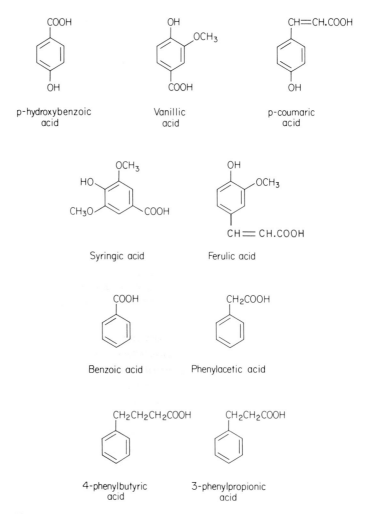

Fig. 7.6   Aromatics in soil.

concentrations in soil. A particular problem in the analysis of aromatics is that they do not appear to accumulate in soil as free acids
but instead become polymerized into humic acid (Fig. 7.7), the
material which can be extracted from the soil with weak alkali and
precipitated as a black, amorphous product with acid. During this
treatment, some of the phenolics may be released in the free state.
Humic acid cannot be extracted with cold alkali, whereas fulvic acid
is soluble in both acid and alkali. The polymers are mainly derived
from p-hydroxybenzaldehyde, vanillin and syringaldehyde.

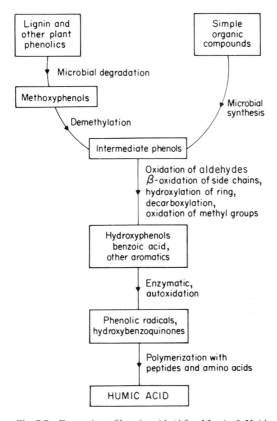

Fig. 7.7 Formation of humic acid. (After Martin & Haider [10].)

Antibiotics

In the 1950s, the ecological significance of antibiotic production by
micro-organisms in soil was investigated by Brian [3]. It was recognized that whereas antibiotics could be considered as inhibitors of
plant growth they could sometimes stimulate it, either directly or by
the removal of pathogens which are inhibitory to growth. Antibiotics
are often metabolic inhibitors of great specificity and potency and
amongst those with inhibitory effects at low concentrations (5 μg
ml^{-1} or less) are actidione, azaserine, alternaric acid and polymyxin.
However, Brian concluded that the diverse observations did not
establish any ecological significance for antibiotics in soil, conclusions which have not been challenged since.

A possible exception to this is where ample substrates, such as
plant residues, are present in soil and localized production of
antibiotics can occur. In the stubble-mulch practice (section 9.2)

where large amounts of straw accumulate, it was [20] concluded that the antibiotic and mycotoxin—patulin—was responsible for the observed phytotoxicity:

However, the authors did not show that it actually accumulated in phytotoxic concentrations in soil and more recent studies have been unable to detect it in soil or straw. Most of the studies had concerned patulin production by *Penicillium urticae*. As relatively few *Penicillium* species have this property, it seems that their populations might build up only in exceptional circumstances.

7.3.3   *Hydrogen sulphide*

Hydrogen sulphide is produced by obligate anaerobic bacteria if there are abundant sulphate and organic substrates at redox potentials below zero. These bacteria are unable to use oxygen and require sulphate as a terminal electron acceptor. The genera are *Desulfotomaculum, Desulfovibrio, Desulfomonas* and *Desulfuromonas*.

Sulphide production has been observed in the rhizosphere [7] and in the spermosphere [6], including the rice rhizosphere where oxygen diffusing out of roots would be expected to produce a locally oxidizing atmosphere. In conditions where sulphide persists, it can exceed concentrations (*c.* $0.1 \mu g$ ml^{-1} water) that inhibit respiration and poison cytochrome oxidase in rice roots. Damage to rice shoots is known in different parts of the world as browning, brusome, akiochi, hie-imochi, suffocation and straight-head diseases. This has been recognized particularly in Asia and Hungary where the necessary conditions prevail. Control of the disease is best achieved by preventing sulphide formation through increasing the soil redox potential by the application of nitrate, rather than by promoting the binding of sulphide. Alternatively, when the catalase activity of rice roots can be stimulated, peroxidases accumulate and promote the activity of *Beggiatoa*, an organism which oxidizes sulphide [21].

7.4   **Conclusion**

Of the great range of microbial products, few have satisfied the conditions necessary to be classified as significant soil phytotoxins or plant growth regulators. Perhaps the short-chain aliphatic acids and hydrogen sulphide are amongst the best authenticated examples of significance to plants. Our searches for the compounds should continue, especially with modern instrumental techniques, because some of them may have potential as agrochemicals to promote plant growth or act as herbicides. There must surely be compounds from specific organisms of relevance; phytotoxins from specific pathogens are recognized so why not from subclinical pathogens?

References

1 Abeles F. B. (1973) *Ethylene in Plant Biology*. Academic Press, New York.

2 Austin D. J., Bu'lock J. D. & Gooday G. W. (1969) Trisporic acids: sexual hormones from *Mucor mucedo* and *Blakeslea trispora*. *Nature (London)*, **223**, 1178-9.

3 Brian P. W. (1957) The ecological significance of antibiotic production. In *Microbial Ecology*, eds Williams R. E. O. & Spicer C. C., pp. 168-88. Cambridge University Press, Cambridge.

4 Brown M. E. (1974) Seed and root bacterization. *Annual Review of Phytopathology*, **12**, 181-97.

5 Chandrasekaran S. & Yoshida T. (1973) Effects of organic acid transformations in submerged soils on growth of the rice plant. *Soil Science and Plant Nutrition*, **19**, 39-45.

6 Dommergues Y. & Jacq V. (1972) Microbiological transformations of sulphur in the rhizosphere and spermosphere. *Annales agronomiques*, **23**, 201-15.

7 Ford H. W. (1973) Levels of hydrogen sulfide toxic to citrus roots. *Journal of the American Society of Horticultural Science*, **98**, 66-8.

8 Hammence J. H. (1946) The determination of auxins in soils. *The Analyst*, 71, 111-16.

9 Ko W. H., Hora F. K. & Herlicksa E. (1974) Isolation and identification of a volatile fungistatic substance from alkaline soil. *Phytopathology*, **64**, 1398-1400.

10 Lockwood J. L. & Filonow A. B. (1981) Responses of fungi to nutrient-limiting conditions and to inhibitory substances in natural habitats. *Advances in Microbial Ecology*, **5**, 1-61.

11 Lynch J. M. (1972) Identification of substrates and isolation of micro-organisms responsible for ethylene production in the soil. *Nature (London)*, **240**, 45-6.

12 Lynch J. M. (1976) Products of soil micro-organisms in relation to plant growth. *CRC Critical Reviews in Microbiology*, **5**, 67-107.

13 Lynch J. M. (1980) Effects of organic acids on the germination of seeds and growth of seedlings. *Plant, Cell and Environment*, **3**, 255-9.

14 Lynch J. M. (1982) Limits to microbial growth in soil. *Journal of General Microbiology*, **128**, 405-10.

15 Lynch J. M., Gunn K. B. & Panting L. M. (1980) On the concentration of acetic acid in straw and soil. *Plant and Soil*, **56**, 93-8.

16 Lynch J. M. & Harper S. H. T. (1974) Formation of ethylene by a soil fungus. *Journal of General Microbiology*, **80**, 187-95.

17 Lynch J. M. & Harper S. H. T. (1980) Role of substrates and anoxia in the accumulation of soil ethylene. *Soil Biology and Biochemistry*, **12**, 363-7.

18 Lynch J. M. & White N. (1977) Effects of some non-pathogenic micro-organisms on the growth of gnotobiotic barley plants. *Plant and Soil*, **47**, 161-70.

19 Martin J. P. & Haider K. (1971) Microbial activity in relation to soil humus formation. *Soil Science*, **111**, 54-63.

20 Norstadt F. A. & McCalla T. M. (1963) Phytotoxic substances from a species of *Penicillium*. *Science*, **140**, 410–11.

21 Pitts G., Allan A. I. & Hollis J. P. (1972) *Beggiatoa*: occurrence in the rice rhizosphere. *Science*, **178**, 990–2.

22 Primrose S. B. (1976) Ethylene-forming bacteria from soil and water. *Journal of General Microbiology*, **97**, 343–6.

23 Rice E. L. (1974) *Allelopathy*. Academic Press, New York.

24 Schreiner O. & Skinner J. J. (1910) *Some Effects of a Harmful Organic Soil Constituent*. US Department of Agriculture Bulletin No. 70, Washington DC.

25 Smith A. M. & Cook R. J. (1974) Implications of ethylene production by bacteria for biological balance of soil. *Nature* (*London*), **252,** 703–5.

26 Smith K. A. & Restall S. W. F. (1971) The occurrence of ethylene in anaerobic soil. *Journal of Soil Science*, **22**, 430–43.

27 Smith K. A. & Robertson, P. D. (1971) Effect of ethylene on root extension of cereals. *Nature* (*London*), **222**, 769–71.

28 Smith K. A. & Russell R. S. (1969) Occurrence of ethylene and its significance in anaerobic soil. *Nature* (*London*), **222,** 769–71.

29 Van Andel O. M. & Fuchs A. (1972) Interference with plant growth regulation by microbial metabolites. In *Phytotoxins in Plant Disease*, eds Wood R. K. S., Ballio A. & Granti A., pp. 227–49. Academic Press, London.

30 Wainright M. & Pugh G. J. F. (1975) Phenol auxins and Ehrlich reactors in soils. *Soil Biology and Biochemistry*, **7**, 287–9.

The simplest and most lumpish fungus has a peculiar interest to us, compared to a mere mass of earth, because it is so obviously organic and related to ourselves, however remote.

H. D. Thoréau in entry for 10th October 1858. *Journal*, 1903.

The intensive use of chemicals in modern agriculture and the generation of large quantities of waste from animal production units has aroused public concern as to how soil amendments might affect man and the environment. Even land-loading of animal waste and heavy application of manures, which have been used for centuries, have caused concern. In the USA, government monitoring and control of agrochemical usage is by the Environmental Protection Agency and similar bodies exist in other countries. The standards set are generally very high and, in addition, most major agrochemical companies spend very large sums of money on their own controls. Yet it seems that the environmentalists are seldom satisfied and their arguments are sometimes highly emotive, as in *Silent Spring* [6]. Is their concern justified? The microbiologist is often at the forefront of such assessment and the following serve as a few brief examples of situations in which their analyses are as essential as the arbitrator.

8.1   Pesticides

In the Western world there has been a great expansion in the use of pesticides, particularly herbicides, for agriculture. As environmental problems have been identified, some compounds have been withdrawn from the market. One classic example is that of DDT. In comparison, in China pesticide usage is not great but DDT is still used fairly widely. It is banned in most countries, despite its benefits being much greater than its harmful side-effects.

Herbicide interactions with the soil microflora have been investigated more extensively than those of fungicides and insecticides but the concepts for study are similar [14].

The microbiologist's concern is twofold. First, are any of the compounds so resistant to degradation by micro-organisms (i.e. recalcitrant) that they become 'bioaccumulated' or 'biomagnified'? Second, does the compound affect any of the useful microbiological activities in soil, thereby affecting soil fertility? All the earlier chapters of this book are therefore relevant to the study of pesticide effects. In Chapter 9, some possible iatrogenic (section 6.4) and other side-effects are considered.

Degradation

The herbicide 2,4-D (dichlorophenoxy acetic acid) is extensively used in most countries and it has been shown that in most soils the compound is readily degraded by several genera of soil bacteria. The enzymes have been characterized and there are about eight biochemical steps with the following result:

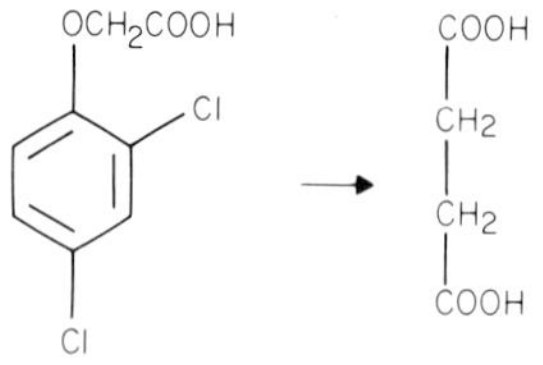

2,4-D

By contrast, 2,4,5-T (trichlorophenoxy acetic acid) is degraded only with very great difficulty and is considered to be recalcitrant:

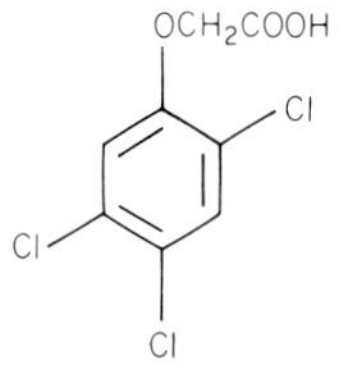

2,4,5-T

It is now regarded as undesirable and has been removed from most markets due to dioxin contamination and its effects on man.

It has been demonstrated that some pesticides are degraded directly, usually by induced enzymes, whereas others are broken down by the common alternative route of co-oxidation or co-metabolism (Table 8.1). It is also now clear that we cannot necessarily expect xenobiotic compounds (chemicals foreign to the natural environment) to be degraded by pure cultures; microbial communities, producing enzymes co-operatively, are the most likely sources of degrading power [3]. There is also some evidence for the formation of adaptive enzymes by soil micro-organisms to deal with otherwise recalcitrant molecules by the so-called 'enrichment effect'. This effect is not exclusive to recalcitrant chemicals. A major problem, however, is that data are often obtained only from laboratory experiments and cannot be extrapolated directly to field conditions. A further note of caution is that the degradation products may themselves be more persistent (Fig. 8.1) or more toxic to animals, plants or micro-organisms than the original chemicals.

Table 8.1 Classification of microbial activities in connection with the decomposition of herbicides. (After Torstensson [15].)

A. Decomposition to smaller molecules
   1. Enzymatic reactions
      (a) Direct degradation of herbicides in the central metabolism of micro-organisms in which they serve as energy sources to supply growth (catabolism) and where adaptation phenomena appear
      (b) Incidental transformation of herbicides by micro-organisms, via peripherical metabolic processes, in the absence of the perfect co-ordination of the process which is characteristic in central metabolism (cometabolism)
      (c) Incidental transformation of herbicides by extracellular enzymes
   2. Non-enzymatic reactions
      (a) Contribution through pH changes
      (b) Production of substances that interact with herbicides in photochemical and chemical reactions
B. Formation of new substances in which herbicides or metabolites of herbicides are parts

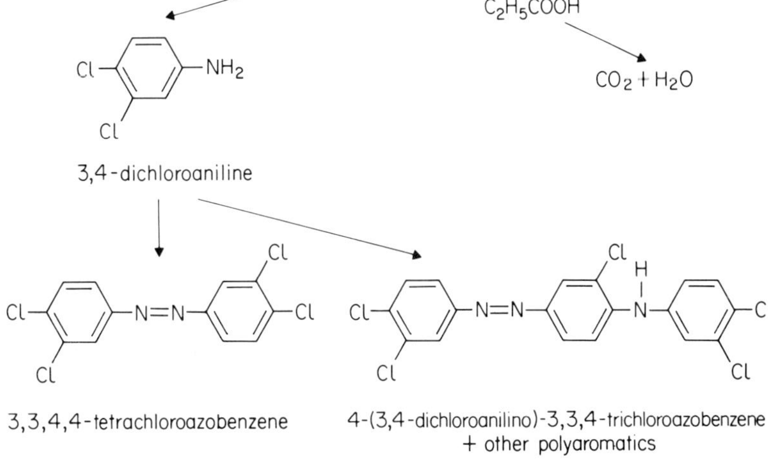

N-(3,4-dichlorophenyl)-propionamide (propanil)

C_2H_5COOH

$CO_2 + H_2O$

3,4-dichloroaniline

3,3,4,4-tetrachloroazobenzene

4-(3,4-dichloroanilino)-3,3,4-trichloroazobenzene + other polyaromatics

Fig. 8.1 Degradation of the herbicide propanil. The aromatic part of the molecules is dimerized and polymerized to persistent residues or may bind with humic complexes to give only very slow degradation.

8.1.2   *Reasons for recalcitrance*

Recalcitrance of chemicals may not be an absolute barrier. There are many factors which contribute to recalcitrance such as: (1) absence of water or oxygen; (2) presence of toxins, such as acetic acid; (3) low temperature and sometimes high hydrostatic pressure; (4) chemical combination so that substrates are immobilized; (5) lack of access spatially; and (6) low substrate concentration. When these barriers can be overcome, recalcitrance is sometimes alleviated.

8.1.3   *Side-effects*

All the methods so far discussed of measuring microbial populations, such as counts and biomass measurements, have been used to study the effects of pesticides on soil. Occasionally, microcalorimetry (section 3.2) has been applied. The activity of the microflora, as measured by respiration, and the potential activity, measured indirectly in enzyme assays, have commonly been determined. The chemical determination of products from metabolic processes during nitrogen transformations and organic-matter decomposition are also particularly relevant.

It is important to recognize that environmental concern arises from the use of pesticides on agricultural crops which enter the food chain. As plants grow in the soil, it is perhaps surprising that most investigators do not consider the effect of plant roots and their

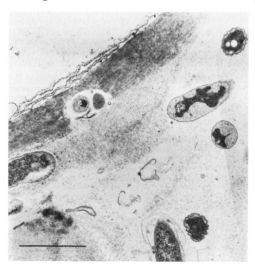

Fig. 8.2 Bacteria penetrating the fourth layer of cortical cells of wheat roots after foliar treatment with the herbicide mecoprop. Bar marker = 1 μm. (Photograph by M. P. Greaves & J. A. Sargent, ARC Weed Research Organization.)

associated rhizosphere when determining pesticide effects. The case for assessing effects on the rhizosphere has been argued eloquently by Greaves *et al.* [12,13]. It has been demonstrated that the herbicide mecoprop can increase the chances of endorhizosphere colonization so that the cortex decays prematurely and plant growth may be reduced (Fig. 8.2). In assessing effects on the rhizosphere, both symbiotic and non-symbiotic organisms should be considered.

Other non-target effects of pesticides include their potential role in increasing plant diseases [2], e.g. some soil-applied herbicides increase the leakage rate of metabolites from roots, so providing substrates for pathogens. About 24 diseases with 19 different vectors are known to be increased by herbicides and there may be others which have not been investigated.

In assessing the effects of pesticides on the soil microflora, it is essential that attention be paid to the applied dose. Even though massive overdoses (tenfold) may induce effects, recommended doses may have no effect. No laboratory investigations should be conducted without including the recommended doses but equally no laboratory investigation should be done without an overdose as a safety factor.

Silage effluent

Silage, which is widely used for animal feed, is prepared by the bacterial fermentation of green-plant tissue. During the fermentation, organic acids are formed and these stabilize the material, preventing further decay and thus allowing long-term storage. Microbiologically, the effluent composition is similar to the silage itself, in that it contains lactic acid, spore-bearing bacteria, yeasts and other fungi. The proportion of lactic-acid producers increases during the fermentation. The effluent from silage represents a dry matter loss of about 10 per cent and contains plant juices which promote microbial growth [16]. This requires oxygen, creating a biological oxygen demand (BOD) (Table 8.2). Thus, if the effluent finds its way into

Table 8.2 Biological oxygen demands of waste. (The figures are representative values but can vary greatly depending on the source of the waste.) (After Woolford [16].)

	BOD (mg O_2 l^{-1})
Silage effluent	90000
Pig slurry	35000
Cow urine	19000
Cow slurry	5000
Domestic sewage	500

watercourses, it can result in the death of fish and other aquatic fauna through the promotion of deoxygenation of the water. The BOD could also deplete soil oxygen concentrations and the presence of organic acids could result in phytotoxicity (section 9.3) but there has been little study of this. Scorching of grassland by the effluent has been observed on occasions and it is now recommended that the effluent be diluted before application to land or that lime be added. Silage effluent has a fertilizer value: at $137 \ m^3 \ ha^{-1}$, the nitrogen, phosphorus and potassium values are 320, 44 and $480 kg \ ha^{-1}$ respectively. Although there is limited production of the material, its application needs careful management to get maximum positive effects and minimum inhibitory effects on plants. Biological treatment of the effluent would have to be cheap to be economic and some success has been achieved with *Arthrobacter* sp. and *Torulopsis utilis*.

8.3   **Application of animal waste to land**

Like silage, effluent animal slurries or feedlot waste are rich sources of organic substrates for the soil microflora. As such, the analysis of their decomposition can be treated in the same way as the decomposition of straw (Chapter 4). The important differences to recognize are that the carbon/nitrogen ratio of animal waste is much smaller and both the carbon and nitrogen are relatively more available [10]. However, their availability in animal waste is dependent on its origin and how it was stored. Applications to land as fertilizer, unlike land-loading, are seldom very heavy (Fig. 8.3).

Fig. 8.3 Spreading of manure from a cattle feedlot on land with vines in Coahuila State, Mexico.

With the heavy applications of waste to land (land-loading), the emphasis is on disposal of the material rather than its potential usefulness for modifying the soil environment. Thus, the retention time, or half-life, of the material is usually of primary concern. Inevitably, the physical environment of the soil (Chapter 2), particularly water availability and temperature, affects this. The slurries must be managed to minimize adverse effects and maximize the fertilizer value to soil. The decreased soil oxygen concentration which can result will cause the redox potential to be decreased, with consequent effects on soil metabolism (Table 8.3).

Table 8.3   Effect of redox potentials on soil metabolism.

Oxygen status	Metabolism	E_h volts	Organic acids
Stage 1 Aerobic (rapid)/microaero-philic	O_2 respiration* NO_3^- reduction* Mn^{4+} Fe^{3+} reduction	+0.6 to +0.3	Only accumulate if fresh organic matter present
Stage 2 Anaerobic (slow)	SO_4^- reduction CH_4 production*	0 to −0.22	Rapid accumulation in early phase, rapid decrease later

* Metabolic processes particularly relevant to the application of waste to land.

An ideal opportunity to study the effects of manure in comparison with nitrogen, phosphorus and potassium fertilizers was available at the Askov Experimental Station in Denmark, where unmanured, manured and fertilized plots had been treated in the same way since 1893 [8]. Until 1973 liquid and solid manure were used but since then only liquid has been used. The manure and the fertilizers utilized per hectare had the same composition of plant nutrients (112·5 kg nitrogen, 18 kg phosphorus and 112·5 kg potassium). Taking an average of all soil samplings, plate counts of fungi, ATP content, oxygen uptake and dehydrogenase activity yielded the highest results for the fertilized soil, lower results for the manured soil and the lowest results for the unmanured soil (Fig. 8.4). Plate counts of bacteria and biomass, determined by the fumigation method (section 3.2), were greatest in the manured soil. However, in general, the differences were small and sometimes insignificant. This suggests that the input of organic matter from manure to the overall soil organic matter is rather small by comparison with the input from plants.

The composition of the soil atmosphere beneath feedlot waste and cropped soil have been compared [9] (Table 8.4). The results of anaerobic metabolism were evident throughout the year. Little

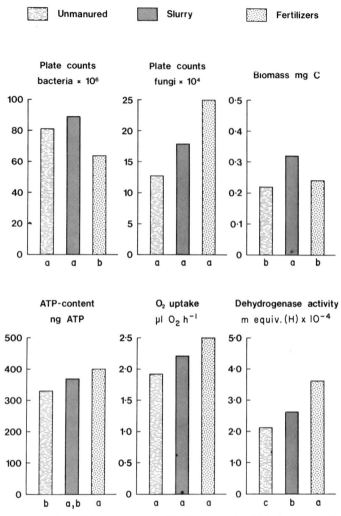

Fig. 8.4 Effects of slurry manure and fertilizers on microbiological properties of soil at Askov Experimental Station, Denmark. The values with the same letter are not significantly different ($P = 0.05$). (From Eiland [8].)

nitrate-nitrogen was leached into the profile of the feedlot, probably because it was denitrified, thus producing little environmental hazard.

Farm livestock wastes (slurries) in Europe are usually produced in smaller quantities than wastes from the feedlots of North America, where animals are stocked more intensively, and disposal is commonly by land-loading. Burford [4] studied the effect of single applications of slurry to grassland and showed that there was an effect for a few months (Fig. 8.5). The soil never became fully

Table 8.4 Average soil atmosphere composition at various soil depths beneath a feedlot and a cropped field for a one-year period. (From Elliott & McCalla [9].)

Depth (cm)	Gas composition (% by volume)				Total detected
	CO_2	O_2	N_2	CH_4	
Feedlot					
31	15·0	9·5	66·0	8·0	98·5
46	12·5	9·5	68·5	8·0	98·5
61	15·0	7·5	65·0	11·5	99·0
76	23·0	4·5	45·0	27·5	100·0
91	18·5	5·0	58·5	18·0	100·0
107	18·0	5·0	57·5	17·5	98·0
122	18·5	5·0	56·5	18·5	98·5
152	21·0	6·0	52·0	20·0	99·0
Cropped field					
31	2·0	18·0	79·0	0	99·0
61	2·0	18·0	80·0	0	100·0
91	2·5	18·0	79·0	0	99·5
122	2·5	18·0	79·0	0	99·5
152	1·0	19·0	79·0	0	99·0

anaerobic and pasture establishment was possible within a few months. The wastes contain organic acids from the microbial fermentation of cellulose in the rumen and this and subsequent fermentation by the soil microflora could cause phytotoxic effects in the periods immediately after application to the land.

The pollution of groundwater is a potential problem with improper management, as the waste provides a rich source of nitrogen so that if nitrification is substantial nitrate will leach down the profile (Fig. 8.6). In winter, on a well-managed grassland, about 60 mg N 1^{-1} was found in drainage water [5] but only 6 per cent of added nitrogen entered this water. This amount does not normally appear to present a great environmental hazard, such as in the eutrophication of water or when babies drink the purified water.

The microbial composition of the slurries has been studied [1,11]. As is to be expected from the chemical evidence, the populations of nitrate-reducing bacteria and, to a lesser extent, nitrite-reducers and urea-hydrolysers increased following slurry application. After 17 weeks, the size of populations declined. Following application of slurries to land, fungal growth decreased initially, then increased over a period of four to eleven weeks and then again decreased. Colony counts of aerobic heterotrophic eubacteria and mesophilic streptomycetes *c.* 10 cm beneath the surface increased after five to six weeks and then decreased to levels similar to those recorded before

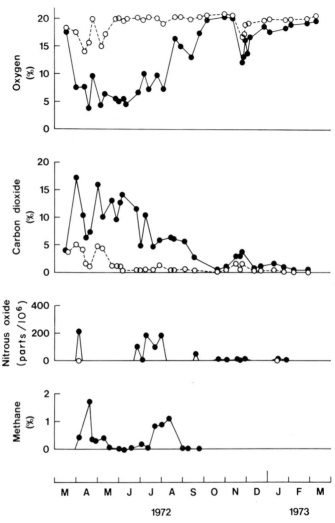

Fig. 8.5 Soil atmosphere under slurry-treated (solid circle) and untreated (open circle) grassland at a depth of 10 cm. Percentages are v/v. (After Burford [4].)

application. A secondary increase occurred after 14 weeks and again declined. In plots receiving the largest amounts of slurry (80 t ha^{-1}), the increases were about tenfold the control (untreated soil).

The microbial populations in soil receiving large applications of cattle feedlot waste have also been studied [7]. Various groups of soil organisms were counted (fungi, aerobic and anaerobic bacteria, *E. coli* types, nitrifiers and denitrifiers). Although numbers increased due to the treatment, the increases were very small and did not persist. Whereas counts are generally less useful than biomass

NITRATE ACCUMULATION

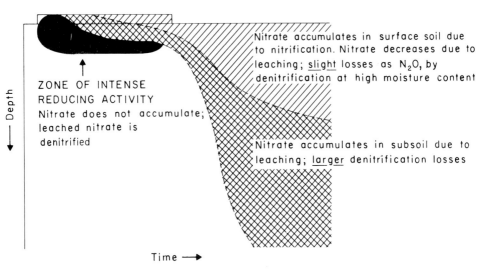

Depth

ZONE OF INTENSE
REDUCING ACTIVITY
Nitrate does not accumulate;
leached nitrate is
denitrified

Nitrate accumulates in surface soil due
to nitrification. Nitrate decreases due to
leaching; slight losses as N_2O, by
denitrification at high moisture content

Nitrate accumulates in subsoil due to
leaching; larger denitrification losses

Time ⟶

Fig. 8.6   Pattern of nitrate accumulation and loss following a heavy slurry application. (After Burford *et al.* [5].)

measurements, the general conclusion that feedlot waste does not deleteriously affect soil micro-organisms seems justified.

8.4   **Conclusion**

Any material, natural or synthetic, applied to soil can affect microbiological activity and soil fertility. Provided the applications are managed properly, harmful effects on plant growth and the environment can usually be avoided. It must also be considered that such material can often be potentially beneficial to soil structure and plant growth. Research effort should be directed at understanding and harnessing these beneficial effects with maximum efficiency.

References

1 Agricultural Research Council (1976) *Studies on Farm Livestock Wastes*. ARC, London.
2 Altman J. & Campbell C. L. (1977) Effect of herbicides on plant diseases. *Annual Review of Phytopathology*, **15**, 361–85.
3 Bull A. T. (1980) Biodegradation: some attitudes and strategies of micro-organisms and microbiologists. In *Contemporary Microbial Ecology*, eds Ellwood D.C., Hedger J.N., Latham M.J., Lynch J.M. & Slater J.H., pp. 107–36. Academic Press, London.

4 Burford J.R. (1976) Effect of the application of cow slurry to grassland on the composition of the soil atmosphere. *Journal of the Science of Food and Agriculture*, **27,** 115–36.

5 Burford J.R., Greenland D.J. & Pain B.F. (1976) Effect of heavy dressings of slurry and inorganic fertilizers applied to grassland on the composition of drainage waters and the soil atmosphere. In *Agriculture and Water Quality*, MAFF Technical Bulletin No. 32, pp. 432–43. HMSO, London.

6 Carson R. (1963) *Silent Spring*. Hamish Hamilton, London.

7 Davis R.J., Mathers A.C. & Stewart B.A. (1980) Microbial populations in Pullman clay receiving large applications of cattle feedlot waste. *Soil Biology and Biochemistry*, **12,** 119–24.

8 Eiland F. (1980) The effects of manure and NPK fertilizers on the soil microorganisms in a Danish long-term field experiment. *Danish Journal of Plant and Soil Science*, **84,** 447–54.

9 Elliott L.F. & McCalla T.M. (1972) The composition of the soil atmosphere beneath a beef cattle feedlot and a cropped field. *Soil Science Society of America Proceedings*, **36,** 68–70.

10 Gilmour C.M., Broadbent F.M. & Beck S.M. (1977) Recycling of carbon and nitrogen through land disposal of various wastes. In *Soils for Management of Organic Wastes and Waste Waters*, eds Elliott L.F. & Stevenson F.J., pp. 173–94. American Society of Agronomy, Madison.

11 Grainger J.M. & Varnam A.H. (1974) Some bacteriological changes associated with the application of cow slurry to grassland. *Journal of Applied Bacteriology*, **37,** viii–ix.

12 Greaves M.P., Davies H.A., Marsh J.P. & Wingfield G.I. (1976) Herbicides and soil micro-organisms. *CRC Critical Reviews in Microbiology*, **5,** 1–38.

13 Greaves M.P. & Malkomes H.P. (1980) Effects on soil microflora. In *Interactions Between Herbicides and the Soil*, ed. Hance R.J., pp. 223–53. Academic Press, London.

14 Hill I.R. & Wright S.J.L. (eds) (1978) *Pesticide Microbiology*. Academic Press, London.

15 Torstensson L. (1980) Role of micro-organisms in decomposition. In *Interactions Between Herbicides and the Soil*, ed. Hance R.J., pp. 159–78. Academic Press, London.

16 Woolford M.K. (1972) The problem of silage effluent. *Herbage Abstracts*, **48,** 397–403.

MICRO-ORGANISMS IN AGRICULTURAL SYSTEMS

The prosperity and happiness of a large and populous nation depend: (1) upon the division of land into small parcels, (2) upon the education of the proprietors of the soil.

Justin Morrill (Vermont) Congressional speech, 1858.

To understand the role of micro-organisms in agriculture and to utilize them, it is essential to conduct complementary field and laboratory studies. Some examples of the application of theoretical and practical work, described in earlier chapters, to the analysis of different aspects of agriculture and horticulture are described in this chapter.

9.1   Effects of reduced cultivation on the soil biomass

Until recently it was considered essential to plough (till) soil before seeding a new crop. A major function of tillage was weed control but the availability of modern herbicides has obviated this reason in many areas. Many soils are now suitable for direct drilling or reduced cultivation and yields are better than or as good as with conventional tillage. A major advantage in soils which are subject to erosion is that residues from the previous crop can be left on the soil surface. In other situations where straw is burnt, the major advantage can be timeliness of cultivation, sometimes making the difference between seed being sown in the autumn (taking advantage of the higher-yielding winter crops) and having to wait until the spring. The effect of direct drilling or reduced cultivation on soil properties and root growth has been investigated but only recently have the effects on the soil microflora been considered.

Using the fumigation-respiration technique of biomass determination (section 3.2), it was shown that compared with conventional seeding following ploughing, direct drilling resulted in a larger microbial biomass in the surface 5 cm of soil. However, this was only evident in the summer when there was a greater root biomass in this horizon (Table 9.1). Thus, it seems that the greater biomass results from the rhizosphere effect. Direct drilling often requires more fertilizer nitrogen to be added to soil than conventional seeding. This can either be the cause or the consequence of the larger biomass in direct-drilled soils [22]. When straw residues are left on the soil surface, the soil biomass may also be greater than when the residues are burnt but this does not appear to be sustained [22].

Biomass only measures gross changes in the microflora. Doran [4,5] measured the changes in specific microbial groups at various locations in the USA. His results are summarized in Tables 9.2 and

Table 9.1 Effect of cultivation on the soil biomass in the surface 5 cm at different times in the growing season of winter wheat sown on 10th October. (After Lynch & Panting [21].)

Date	Cultivation	Biomass (mg C $100 g^{-1}$ dry soil)	% soil organic carbon in biomass
1977			
2 August	Direct drilled	72*	1·6
	Ploughed	54	1·3
24 November	Direct drilled	51	1·2
	Ploughed	66	1·3
1978			
27 April	Direct drilled	74	1·5
	Ploughed	51	1·1
22 June	Direct drilled	122	2·7
	Ploughed	106	2·6
29 June	Direct drilled	121	2·6
	Ploughed	107	2·6
24 August	Direct drilled	121*	2·0
	Ploughed	71	1·4

* Significantly different from ploughed ($P < 0.05$).

9.3. It should be emphasized that the zero tillage system in the USA has straw residues present as standard (section 9.2). In much of Britain, residues are disposed of by burning or baling. Nevertheless, the results from the USA showing that most microbial groups increase under zero tillage (Table 9.2) are consistent with the biomass data collected in England. The phosphatase and dehydrogenase levels were also greater. Aerobic nitrifiers were greater with tillage at six of the seven locations. The presence of more soil water appeared to

Table 9.2 Microbial counts and other changes in the surface 7·5 cm of soil. The results are the means for seven locations in the USA. (After Doran [5].)

Measurement	Ratio untilled/ploughed
Total aerobes	1·35
Fungi	1·57
Actinomycetes	1·14
Aerobic bacteria	1·41
NH_4^+ oxidizers	1·25
NO_2^- oxidizers	1·58
Facultative anaerobes	1·57
Denitrifiers	7·31
Potentially mineralizable nitrogen	1·35
Total organic carbon	1·25
Kjeldahl nitrogen	1·20
Soil-water content	1·47

the author to be the major factor responsible for the changes but it seems likely that soil pH, organic carbon and nitrogen may also regulate them. The trends were less marked or reversed at lower depths.

Again consistent with the biomass data, microbial populations increase in the presence of straw residues (Table 9.3) and, with the exception of *Nitrosomonas*, the increase in all groups correlated with the residue application rate. The pH in the presence of residues (5.26) was greater than without residues (5.15). The increases in microbial populations were probably due to both the increased soil-water content and the extra available substrates.

Table 9.3   Ratio of microbial counts on 14th June 1977 in soil (0–7·5 cm deep) where surface corn residues (14 t ha^{-1}) were left mainly between rows, compared with soil receiving no residues. (After Doran [4].)

Measurement	Ratio residues present/residues not present
Total aerobes	5·0
Bacteria	6·1
Actinomycetes	2·4
Fungi	1·9
Denitrifiers	43·7
Nitrosomonas	2·4
Nitrobacter	3·9

9.2   **Problems from stubble mulch and conservation tillage**

Stubble mulch, a method in which wet crop residues are left on the soil surface, has been practised in the Midwest of the USA since the 'dustbowl era', with the aim of controlling wind erosion. With this tillage method, crop residues are managed on the soil surface as much as possible but it became apparent in the 1950s that crop yields were reduced by the microbial decomposition of the residues [24]. Patulin appeared to be the phytotoxin involved but recent work has not substantiated this (section 7.3).

The problem arose again when conservation tillage systems were introduced into the Pacific North-west of the USA [7]. This region produces some of the greatest dry-land, winter-wheat yields in the USA but because the soils are generally unstable and most fields are on sloping hillsides, it is desirable to manage crop residues on the soil surface to prevent water (gully, sheet, rill, slip) and wind erosion (Fig. 9.1). In some years, poor wheat stands result and plants from high-residue areas are much smaller than those from low-residue areas (Fig. 9.2). Raised crowns are quite characteristic of the problem

Fig. 9.1 Gully (water) erosion on sloping hillsides of the Pacific North-west (near Rosalia, Washington).

Fig. 9.2 Comparison of a winter-wheat plant taken from a low-residue area (left) and a high-residue area (right). (Photograph by L. F. Elliott, Washington State University.)

(Fig. 9.3). Using seedling bioassays, where extracts from the residues decomposing in the field are added to agar containing wheat seedlings with no other carbon source [2], root growth is often inhibited, particularly during the early stages of straw decomposition (Fig. 9.4). Consistent with similar observations in the USSR [25] and in Britain (section 9.3), it appears that the production of short-chain aliphatic acids (acetic, propionic and butyric) are at least partly responsible for the problem [34]. There is synergism between these acids, thereby increasing the damage [35]. The acids have been extracted in phytotoxic concentrations from crop residues decomposing in the field.

Fig. 9.3 Elevated crown node of winter wheat growing in a heavy density of wheat-straw residue. The short, stubby roots do not penetrate the soil. (Photograph by L. F. Elliott, Washington State University.)

The fact that damage does not occur every year has led to searches for other causative factors. Conservation tillage and minimum tillage can increase, decrease or have no effect on plant diseases [33]. The stunted plant growth observed in the field would also be consistent with the effects caused by the pathogen *Pythium*, at least in the low-lying, wetter areas of the fields. Treatment of seed with a fungicide selective against *Pythium* largely overcame the adverse effect of straw at one site [3]. Whereas this may not be a general explanation, it offers an alternative explanation in some situations.

Indeed, another possible interpretation now appears to be that the plant residues also increase and the population of *Pseudomonas* spp., which colonize the endorhizosphere and become subclinical pathogens, also increase [8].

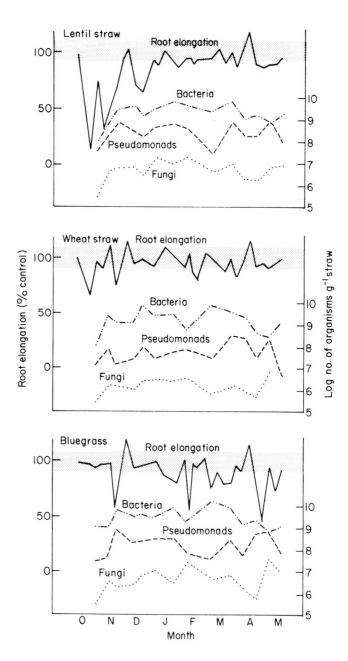

Fig. 9.4 Relative growth of seedling wheat roots in agar medium treated with water extracts of lentil, wheat and bluegrass straw, compared with seedling root growth in agar only. The populations of total bacteria, pseudomonads and fungi are also shown. The shaded bands between 93 per cent and 107 per cent represent the average standard deviation in the control plants. (From Cochran *et al.* [2].)

9.3   **Direct drilling through straw residues in wet autumns**

In Britain it is normal to burn straw residues from the preceding crop. If this is not done and the autumn is wet, establishment of winter cereals can be poor, tiller numbers small and crop yields reduced [19] (Fig. 9.5). Similar data has been obtained by the British Ministry of Agriculture, Fisheries and Food at five experimental farms in different parts of the country over about five years. In dry autumns, the problem is much less.

The problem occurs when straw and seed are forced in close proximity, such as in a smeared drill slit (Fig. 9.6). Anaerobic fermentation of the cellulose in the straw, yields acetic acid in phytotoxic concentrations [17,18] (Fig. 7.5). The acid does not diffuse very far through the straw, its concentration being reduced to one-half within 1·5 cm, but as the sites of production and action are adjacent, this is of little consequence. However, if straw is ploughed in, the amount

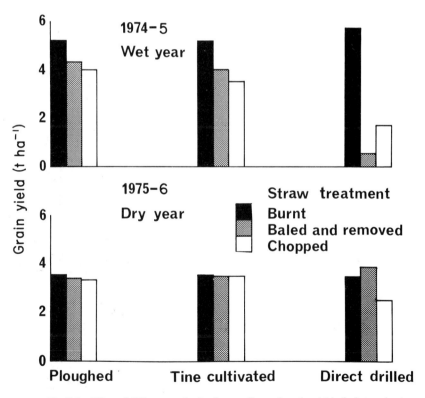

Fig. 9.5   Effect of different methods of straw disposal on the yield of winter wheat established after contrasting methods of cultivation in wet and dry autumns. The results are for the Drayton Experimental Husbandry Farm, near Stratford-upon-Avon.

Fig. 9.6 Smeared drill slit where seed and straw are incorporated in close proximity. Poor establishment of the crop results.

of substrate in the seed-bed is much less and, indeed, ploughed soils seem to suffer relatively little from this syndrome. Although both the water-soluble components and the readily available cellulose and hemicellulose appear to be the critical substrates, they normally disappear from the straw within about a month (section 4.2). If wet weather occurs, a delay in drilling can sometimes minimize the problem. If the early autumn is dry and rainfall then occurs immediately after drilling, resulting in a flush of fermentative activity, then maximum damage to the establishing crop can occur. In other situations, where the substrates are not degraded in the autumn but instead are retained throughout winter, even though microbial activity is slow at the lower temperature, toxins can be produced at a slow but steady rate throughout the winter. As the plant is also growing only very slowly, it is again vunerable. It is, therefore, rather difficult to predict precisely when damage will occur, although the state of straw decomposition can be readily monitored by chemical assay [11,12].

The decomposition of the straw residues by fungi, which can subsequently colonize the seed and prevent seed germination by oxygen competition (section 2.3), is a potential problem but this has not been demonstrated in the field.

Decomposition of straw, which has a large carbon/nitrogen ratio (about 80:1), can also potentially induce a nitrogen-deficiency problem in the crop (section 4.5). With rapid immobilization of nitrogen on to the straw (Fig. 4.10), it is quite common to see some signs of nitrogen deficiency in the establishing crop. However, the nitrogen is immobilized only temporarily and this can be released in the spring when the crop may have more need for it (Fig. 4.11). Indeed, it could be an advantage to have the nitrogen immobilized in winter to prevent leaching or denitrification losses. The rate of straw decomposition does not appear to be critically governed by exogenous nitrogen but rather the availability of carbon appears to be the greatest limitation on the decomposition process [13] (Fig. 9.7). The role of straw in nitrogen cycling has therefore yet to be fully evaluated. Evaluation will only be possible by the use of ^{15}N tracers [15].

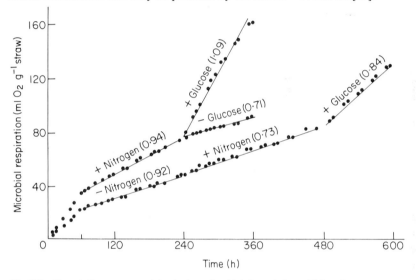

Fig. 9.7   Cumulative oxygen uptake during microbial degradation of 1.5 g wheat straw in 148.5 g sand moistened with 30 ml water. Additions made were 8940 μg nitrogen as $(NH_4)_2SO_4$ or 200 mg glucose. Figures in parentheses are respiratory quotients (CO_2 respired: O_2 uptaken). (From Knapp et al. [13].)

9.4   **Problems of weed residues**

Some of the perennial grass weeds, such as couch (quack) grass, have been difficult to control. The introduction of herbicides like glyphosate (N-phosphonomethyl glycine), which is translocated to the rhizomes, has relieved the problem. However, on a few occasions it has been noticed that when cereal crops are drilled into dense mats of decomposing rhizomes, the establishing plants can sometimes be injured and even die [23,26]. The die-back occurs in the spring after autumn sowing and has been observed most often on light, chalky soils (Fig. 9.8).

The symptoms were observed in detail when growing barley plants in pots of soil in the greenhouse. Plants generally reached the two to

Fig. 9.8 Poor establishment of winter wheat in a circular area of a field where a dense infestation of couch (quack) grass had been successfully killed with the herbicide glyphosate.

three leaf stage, then the oldest leaf became chlorotic, the tip became grey and eventually necrosis was evident in most leaves. The damage was greater in drier soil when the soil-water potential was $-50\,\text{kPa}$ compared with wet soil ($\psi = -17\,\text{kPa}$) and when fertilizer nitrogen was added to the soil. Therefore, as nitrogen is the nutrient most likely to be immobilized by the decomposer micro-organisms, it did not appear that nutrient immobilization explained the damage and, indeed, the leaf symptoms were not consistent with nutrient deficiency.

Like straw, the decomposing rhizomes contain a high proportion of lignocellulose (section 4.1) and therefore production of organic acid toxins provided a likely explanation [20]. Phenolic acids were also produced but in much lower concentrations than acetic and butyric acids. However, as the acids are formed fermentatively they would most likely be produced in the wetter soil. Thus, they could only provide a partial explanation for the observed damage.

Fusarium culmorum was very commonly isolated from the decomposing rhizomes and the plant damage observed would be consistent with symptoms of *Fusarium* foot-rot [28]. Furthermore, low soil-water potentials favour *Fusarium* and the extra fertilizer nitrogen reduces leaf-water potential so that the plant is more susceptible to invasion (section 10.2). Organic acid toxins, which can be produced in anaerobic pockets in otherwise aerated conditions, are produced and probably also stress the plant such that the main damage is done by *Fusarium culmorum*, but the two factors appear to interact (Table 9.4).

Table 9.4   Effect of $5 \, mol \, m^{-3}$ acetic acid and *Fusarium culmorum* on the growth of barley seedlings. (From Penn & Lynch [28].)

	Mean length of first three leaves (mm) 12 days after germination			
Inoculum density (spores ml^{-1})	0	10^4	10^5	10^6
Seedlings treated with				
5 mol m^{-3} acetic acid	120^b	98^c	89^{c,d}	83^{c,d}
No acid treatment	140^a	118^b	114^b	78^d

Results with different letters are significantly different ($P = 0.05$).

No evidence has been found for herbicide residues in the soil. Herbicide accumulates in the rhizomes and can be released on microbial decomposition but it has been shown that this is very unlikely to contribute to the observed damage, especially as it is readily degraded itself by the soil microflora [27]. Thus, the damage can only be regarded as a herbicide side-effect or iatrogenic disease (section 6.4) which can be managed (sections 10.1 and 10.2).

9.5   **Reseeding grassland**

In old pastures which have become unproductive, it is normal to plough them and then reseed. This operation is time-consuming and the farmer is reluctant to keep stock off the land for any longer than is essential. It has therefore become increasingly common to kill off the old sward with herbicide and then to directly reseed through the decomposing herbage. The incidence of failure with this technique is quite high. Sometimes, establishment is so bad that the farmer must reseed yet again.

Microbiologically, we appear to be faced with problems analogous to those outlined in sections 9.3 and 9.4. The grass residues are potential substrates for phytotoxin production under anaerobic conditions (Table 9.5) and it is of interest that two of the species which yield the greatest concentrations of organic acid toxins—creeping

Table 9.5 Acetic acid production and the effect of solutions produced after 20 days of anaerobic decomposition of grasses on root extension. (After Gussin & Lynch [9].)

Residue	Acetic acid produced (mol m^{-3})	Root extension of test species (mm)						
		Alopecurus myosuroides	*Festuca rubra*	*Holcus lanatus*	*Lolium perenne*	*Poa annua*	*Poa trivialis*	*Trifolium repens*
Distilled-water control	0	9	9	10	25	8	6	45
Treatment control	<0·1	10	7	10	32	10	5	48
Agrostis stolonifera	10·5	0	0	3	9	5	0	2
Alopecurus pratensis	13·9	0	5	0	3	0	0	0
Anthoxanthum odoratum	4·5	15	7	14	25	10	6	33
Festuca rubra	16·7	0	0	4	5	0	0	0
Holcus lanatus	0·4	15	9	21	36	12	11	37
Lolium perenne	2·1	6	8	10	26	10	7	22
Poa trivialis	9·9	11	4	7	16	7	3	21
LSD ($P=0·05$)		2	2	2	3	2	2	4

Treatment control contained soil and distilled water only. Common names of species are creeping bent (*A. stolonifera*), black grass or slender fox tail (*A. myosuroides*), common fox tail (*A. pratensis*), sweet vernal grass (*A. odoratum*), creeping red fescue (*F. rubra*), Yorkshire fog (*H. lanatus*), perennial ryegrass (*L. perenne*), annual meadow grass (*P. annua*), rough meadow grass (*P. trivialis*) and white clover (*T. repens*).

red fescue and creeping bent—are among the most difficult residues in which to establish new grass. They are also two of the common grasses in very old pastures. The productive species in Britain that need to be established in swards are perennial rye-grass and white clover. Both are sensitive to the organic acids in the concentration range in which they can be produced from the residues.

Rye-grass is also susceptible to *Fusarium* foot- and root-rots. Both *Fusarium culmorum* (Figs. 9.9 and 9.10) and *F. nivale* can be isolated from the decomposing grasses and they are pathogenic to rye-grass growing in sand moistened with nutrient solution. The pathogenicity

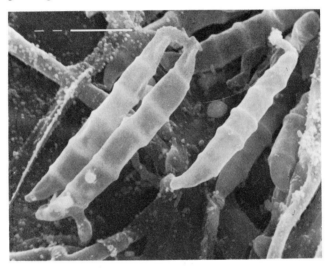

Fig. 9.9 *Fusarium culmorum* which has used grass residues as a substrate base and colonized the seed coat of winter wheat. Note the presence of bacteria on the fungal hyphae and the seed coat. Bar marker = 10 μm. (Photograph by E. J. Gussin and S. F. Young, Letcombe Laboratory.)

of the fungus depends on the substrate base (root residue) present and on the host species. Sometimes infection can be great but the pathogenicity is minimal [10]. There is also the potential for toxins from subclinical pathogens (section 9.2) and pathogens to interact synergistically in causing plant damage and reducing productivity.

Studies in the Netherlands have indicated that a range of fungi and nematodes have the potential to cause the plant damage, indeed the nematodes may be the vectors for the fungi [14]. In pot trials, field trials and samples from problem fields, treatment of samples with nematicides and, to a lesser extent, fungicides reduced the numbers and range of genera of fungal isolates, thereby alleviating the stress on the plant.

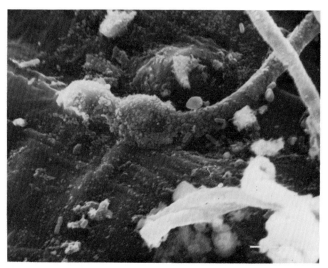

Fig. 9.10 *Fusarium culmorum* colonizing wheat seed and forming an appresorium-like structure. Seed death resulted. Bar marker = 1 μm. (Photograph by E. J. Gussin and S. F. Young, Letcombe Laboratory.)

9.6   Virus disease of red clover

Red clover is now commonly sown in Britain because the tetraploid varieties yield well and contain more protein than grasses. Breeders have produced cultivars resistant to stem eelworm (*Ditylenchus dipsaci*), the main pest of red clover, and to clover rot (section 6.4). A major problem remaining is the occasional occurrence of red clover necrotic mosaic virus (RCNMV) [1] (Fig. 3.2). It causes veinal chlorosis, often followed by severe necrosis, deformation and stunting, reducing yield by more than 50 per cent. Seedlings became infected when grown in pots containing RCNMV-infected plants or soil from infected sites and the roots of infected test seedlings were found to contain the fungus *Olpidium* sp., which may be the vector. *Olpidium* zoosporangia require adequate soil moisture before they release zoospores, so it may be that much infection of spring-sown crops is delayed until the autumn. However, if the disease becomes a major problem, this observation may be a route to the control of the disease (section 10.2). White clover mosaic virus also occurred on red clover but was less damaging. This illustration of a soil virus being responsible for damage to the crop emphasizes that all microbial groups should be considered when diagnosing soil sickness.

Sugar-beet

The sugar-beet seed is very small and prone to infection by *Phoma betae* and damping-off fungi, such as *Pythium*, in the soil. It is critical that all seeds emerge as precision drilling of pelleted seed is now common for the crop and there is, therefore, little scope for compensating for any plants which fail to emerge. Besides pelleting, other seed treatments which are used are washing (to remove germination inhibitors from the seed coat), thiram (fungicide) soaking, advancing (moistening seeds for 24 hours and then drying back) and steeping in ethyl mercuric phosphate. Unfortunately, some of these treatments interact negatively, particularly advancing and steeping, perhaps because both treatments provide a similar function and interactively they may allow mercury to enter the embryo and damage it [16].

In the irrigated areas along the Columbia River in southern Washington state, winter cover crops are used to prevent wind erosion. When winter wheat and oat cover crops were incorporated into the soil in spring and a sugar-beet crop planted, stand losses ranged from 17 per cent to 30 per cent [6]. Phytotoxicity to the sugar-beet seedlings was observed with water, ether and methylene chloride extracts of the cereal residues. In the laboratory, seedlings died when they came into direct contact with decomposing wheat residues taken from the field (Fig. 9.11). Residues from fungicide (captan and pentachloronitrobenzene)-treated plants contained more water-extractable phytotoxin(s), compared with control plants, possibly

Fig. 9.11   Effect of green wheat residues on sugar-beet. Where the sugar-beet roots contacted the wheat piece, the seedlings died. (Photograph by L. F. Elliott, Washington State University.)

because the fungicides inhibited organisms that competed with the toxin-producing organisms. The toxins did not seem to be aliphatic acids because lime did not overcome the toxicity. The damage in the presence of the fungicide makes the sugar-beet fungal pathogen, *Rhizoctonia* sp., seem unlikely to be responsible but *Pythium* sp. could possibly be involved.

9.8   **Replant problems of fruit trees**

When fruit trees are planted on land previously occupied by the same or closely related species, poor growth can result. The symptoms are a small aerial system and poor, often discoloured, roots with diminished laterals and root hairs. This has been recognized for over 250 years as 'soil sickness' or 'replant problems' [36] and although the term 'specific replant disease' has been proposed in order to avoid confusion with the many unrelated replanting problems, this description may not be very useful as trees are not always subject to the problem. Similarly, it is not always necessary to have the same type of tree replanted to observe the problem. It should therefore be regarded as a general soil malaise. It is more severe in establishing apple, cherry, peach and citrus trees and least severe in plum, pear and rose trees. Various options, including pathogens, nutrition, physical and chemical factors, have been considered as possible explanations.

Studies of pot-grown cherry and plum trees suggest the fungus *Thielaviopsis basicola* is responsible [31]. Some strains of the fungus isolated from the soil produced all the features of the disorder, including inter- and intrageneric specificity, host symptoms, establishment, immobility and persistence of the causal agent in soil, normal growth of trees after their transfer to 'non-replant' soils and limited influences of soil type on the incidence of the disorder. Thus, it appears that mature trees can tolerate the presence of the fungus in soil but the establishing trees cannot and that only some strains of the fungus are pathogenic. *T. basicola* does not affect the various species of *Malus* (apple).

The apple problem (Fig. 9.12) was investigated in a similar manner to cherry and plum but proved more intractable [32]. *Pythium sylvaticum* and seven other *Pythium* spp. were isolated from the replant soil and it was found that most of them could reduce tree growth when applied to fresh soil. The depressions in growth were similar to the increases which occurred after chloropicrin fumigation of the replant soil of apple orchards. The fungi had only a low virulence to cherry. The observations are not completely conclusive, however, because poor growth phenomena are much more difficult to diagnose

Fig. 9.12 Growth of apple trees (Cox's Orange Pippin on M9 EMLA root-stocks) two years after planting 12 rows apart. Those on the left were planted on land previously occupied by apple, those on the right were planted on land previously occupied by cherries. (Photograph by G. W. F. Sewell, East Malling Research Station.)

than more definite symptoms, such as lesions, which occur from most pathogens. This is a problem with *Pythium* diseases generally (section 6.4).

9.9   Cavity spot of carrots

The cavity-spot disorder can be initiated anywhere on the carrot storage organ [29,30]. The cells of the outermost layers of the secondary phloem collapse progressively, leading to rupture of the periderm. A wound periderm forms around the lesion (Figs. 9.13 and 9.14). Suberin and lignin are deposited in the cell walls and polyphenols accumulate in the surrounding tissue. The problem is associated with soil compaction and heavy watering but no recognized pathogen is present. The symptomology from the field can be repeated by growing plants in pots of soil. Anaerobic pectolytic bacteria, *Clostridium* sp., have been isolated from the lesions and shown to be pathogenic under wet conditions. When dry conditions return, the lesions resemble those of cavity spot.

These observations are particularly important in demonstrating that anaerobic microsites must exist on the root surface where the anaerobe can grow. Once the pathogen is established, it is possible

Fig. 9.13 Roots of carrot (var. Chantenay) with slight, early symptoms of cavity spot in which the periderm is unbroken. (Photograph by D. A. Perry, Scottish Crops Research Institute.)

Fig. 9.14 Longitudinal slice through a slight lesion in carrot showing the cavity spot; the periderm is intact but the secondary phloem is collapsed. (Photograph by D. A. Perry, Scottish Crops Research Institute.)

that facultative anaerobes and aerobes could use the available oxygen in the soil atmosphere to provide respiratory protection to the anaerobe. This is analogous to the respiratory protection with which nitrogenase is provided (section 6.2).

9.10   **Conclusion**

The significance of modified agricultural practices on the soil micro-flora and nutrient cycling has yet to be fully evaluated but it does suggest that these measurements should be considered alongside those of root growth and soil chemical/physical properties. The few examples in this chapter serve to illustrate that both pathogenic and non-pathogenic (subclinical) micro-organisms have important roles in agricultural productivity and that some problems which have been recognized for a long time by farmers are now explicable. It is clear that the microbiological effect is not the only factor to be considered. The microbiologist must work closely with scientists from other disciplines to build up a scenario which can lead to the eventual diagnosis.

References

1 Bowen R. & Plumb R. G. (1979) The occurrence and effects of red clover necrotic mosaic virus in red clover (*Trifolium pratense*). *Annals of Applied Biology*, **91,** 227–36.

2 Cochran V. L., Elliott L. F. & Papendick R. I. (1977) The production of phytotox-ins from surface crop residues. *Soil Science Society of America Journal*, **41,** 903–8.

3 Cook R. J., Sitton J. W. & Waldher J. T. (1980) Evidence for *Pythium* as a pathogen of direct-drilled wheat in the Pacific North-west. *Plant Disease*, **64,** 102–3.

4 Doran J. W. (1980) Microbial changes associated with residue management with reduced tillage. *Soil Science Society of America Journal*, **44,** 518–24.

5 Doran J. W. (1980) Soil microbial and biochemical changes associated with re-duced tillage. *Soil Science Society of America Journal*, **44,** 765–71.

6 Elliott L. F., Miller D. E. & Richards A. W. (1979) Phytotoxicity of incorporated cover crops to sugar-beet seedlings. *Plant Disease Reporter*, **63,** 882–6.

7 Elliott L. F., McCalla T. M. & Waiss A. (1978) Phytotoxicity associated with residue management. In *Crop Residue Management Systems*, ed. Oschwald W. R., pp. 131–46. American Society of Agronomy, Madison.

8 Elliott L. F. & Lynch J. M. (1983) Pseudomonads as a factor in the growth of winter wheat (*Triticum aestivum L*). *Soil Biology and Biochemistry* (in press).

9 Gussin E. J. & Lynch J. M. (1981) Microbial fermentation of grass residues to organic acids as a factor in the establishment of new grass swards. *New Phytolo-gist*, **89,** 449–57.

10 Gussin E. J. & Lynch J. M. (1983) Infected root residues as substrates for *Fusarium culmorum* on wheat, barley and rye-grass. *Journal of General Microbiology* (in press).

11 Harper S. H. T. & Lynch J. M. (1981) The kinetics of straw decomposition in

relation to its potential to produce the phytotoxin acetic acid. *Journal of Soil Science*, **32**, 627–37.

12 Harper S. H. T. & Lynch J. M. (1981) The chemical components and decomposition of wheat straw leaves, internodes and nodes. *Journal of the Science of Food and Agriculture*, **32**, 1057–62.

13 Knapp E. B., Elliott L. F. & Campbell G. S. (1983) Microbial respiration and growth during the decomposition of straw. *Soil Biology and Biochemistry* (in press).

14 Labruyere R. E. (1979) Resowing problems of old pastures. In *Soil-borne Plant Pathogens*, eds Schippers B. & Gams W., pp. 313–26. Academic Press, London.

15 Ladd J. N. (1981) The use of ^{15}N in following organic matter turnover; with specific reference to rotation systems. *Plant and Soil*, **58**, 401–11.

16 Longden P. C. (1976) Seed treatments to lengthen the sugar-beet growing period. *Annals of Applied Biology*, **83**, 87–92.

17 Lynch J. M. (1977) Phytotoxicity of acetic acid in the anaerobic decomposition of wheat straw. *Journal of Applied Bacteriology*, **42**, 81–7.

18 Lynch J. M. (1978) Production and phytotoxicity of acetic acid in anaerobic soils containing plant residues. *Soil Biology and Biochemistry*, **9**, 305–8.

19 Lynch J. M., Ellis F. B., Harper S. H. T. & Christian D. G. (1980) The effect of straw on the establishment and growth of winter cereals. *Agriculture and Environment*, **5**, 321–8.

20 Lynch J. M., Hall K. C., Anderson H. A. & Hepburn A. (1980) Organic acids from the anaerobic decomposition of *Agropyron repens* rhizomes. *Phytochemistry*, **19**, 1848–9.

21 Lynch J. M. & Panting L. M. (1980) Cultivation and the soil biomass. *Soil Biology and Biochemistry*, **12**, 29–33.

22 Lynch J. M. & Panting L. M. (1982) Effects of season, cultivation and nitrogen fertilizer on the size of the soil microbial biomass. *Journal of the Science of Food and Agriculture*, **33**, 249–52.

23 Lynch J. M. & Penn D. J. (1980) Damage to cereals caused by decaying weed residues. *Journal of the Science of Food and Agriculture*, **31**, 321–4.

24 McCalla T. M. & Norstadt F. A. (1974) Toxicity problems in mulch tillage. *Agriculture and Environment*, **1**, 153–74.

25 Mishustin E. N., Erofeev N. S. & Timiryazev K. A. (1966) Nature of the toxic compounds accumulating during the decomposition of straw in soil. *Microbiology*, **35**, 126–9.

26 Penn D. J. & Lynch J. M. (1981) Effect of decaying couch grass rhizomes on the growth of barley. *Journal of Applied Ecology*, **18**, 669–74.

27 Penn D. J. & Lynch J. M. (1982) Toxicity of glyphosate applied to roots of barley seedlings. *New Phytologist*, **90**, 51–5.

28 Penn D. J. & Lynch J. M. (1982) The effect of bacterial fermentation of couch grass rhizomes and *Fusarium culmorum* on the growth of barley seedlings. *Plant Pathology*, **31**, 39–43.

29 Perry D. A. & Harrison J. G. (1979) Cavity spot of carrots. I. Symptomatology and calcium involvement. *Annals of Applied Biology*, **93**, 101–8.

30 Perry D. A. & Harrison J. G. (1979) Cavity spot of carrots. II. The effect of soil conditions and the role of pectolytic anaerobic bacteria. *Annals of Applied Biology*, **93**, 109–15.

31 Sewell G. W. F. (1981) Effects of *Pythium* species on the growth of apple and their possible causal role in apple replant disease. *Annals of Applied Biology*, **97**, 31–42.

32 Sewell G. W. F. & Wilson J. F. (1975) The role of *Thielaviopsis basicola* in the

specific replant disorders of cherry and plum. *Annals of Applied Biology*, **79,** 149–69.

33 Sumner D. R., Doupnik B. & Boosalis M. G. (1981) Effects of reduced tillage and multiple cropping on plant diseases. *Annual Review of Phytopathology*, **19,** 167–87.

34 Wallace J. M. & Elliott L. F. (1979) Phytotoxins from anaerobically decomposing wheat straw. *Soil Biology and Biochemistry*, **11,** 325–30.

35 Wallace J. M. & Whitehand L. C. (1980) Adverse synergistic effects between acetic, propionic, butyric and valeric acids on the growth of wheat seedling roots. *Soil Biology and Biochemistry*, **12,** 445–6.

36 Worlidge J. (1698) *Systema Agriculturae*. London.

And he gave it for his opinion, that whoever could make two ears of corn, or two blades of grass to grow upon a spot of ground where only one grew before, would deserve better of mankind, and do more essential service to his country, than the whole race of politicians put together.

J. Swift   In Ch. 6 Voyage to Brobdingnag, *Gulliver's Travels*, 1726.

In this concluding chapter we will see that some exciting new developments are taking place in our understanding of the ecology of soil micro-organisms, to the extent that it may soon be possible to manage some of the problems outlined in earlier chapters and, indeed, to optimize the growth of plants by modification of the soil-root environment. The following examples serve to illustrate soil biotechnology in practice.

10.1   Chemicals as control agents

Possibly the simplest way to eliminate soil microbiological problems is to attempt to remove them from the soil by sterilization. Total soil sterilization is impractical of course. A better approach is to use a fumigant such as methyl bromide, chloropicrin or 1,3-dichloropropane which can be pumped from cylinders into the soil under polymer sheets; this provides a partial sterilization. The procedure can be quite successful in eliminating many root diseases, although the target is more commonly nematodes than micro-organisms. However, the cost is somewhat prohibitive and, unless the crop is of particularly high value, such as in the glasshouse industry, it is only really useful as an experimental tool to study root diseases. Heat treatments, green manures and flooding have sometimes been successful in reducing specific diseases. The whole subject of soil disinfestation is reviewed in an extensive monograph [50], which includes sections on the rationale for selection of treatments, the fate of fumigants and biological control methods (sections 10.3 and 10.4).

In the agrochemical industry, there are large numbers of chemicals that have been screened for their activities in protecting plants against diseases, although the number available for the control of root diseases is surprisingly small, e.g. no chemical is available yet for the control of the major wheat disease—take-all. Commonly, chemicals are applied to the seed. Many of them are systemic and will control foliar diseases, such as mildew of wheat. Others are applied to protect the seed and young seedling during establishment. A major problem is that when the seed is placed in the soil, microbial degradation of the chemical can rapidly inactivate it and limit its useful life.

What is the strategy for the selection of a seed dressing or seedling root drench? With cereal seeds, in Britain at least, this is often decided by the seed merchant. The strategy may, in part, be to protect the seed during storage, although this should be quite unnecessary if the seed is stored at the correct humidity and temperature. The soil fauna may also need to be controlled. In practice, many merchants opt for the organo-mercurials, although in some countries they are banned because of their extreme toxicity if they should bioaccumulate in the environment. Gamma BHC is often applied to cereals. This was originally described as an insecticide but it is fortuitous that it also has antifungal action. Liquid dressings, such as triforine, which are applied for their systemic action against fungi can actually promote the growth of saprophytic fungi and increase the microbe's chance of preventing seed germination [40]. Greater protection can be provided by drenching roots rather than seeds in fungicides. This has application particularly in the glasshouse industry, where transplanting is normal. However, very few of the seed dressings are completely satisfactory against the range of pathogens that the seed and establishing seedling may encounter.

The design of the plant protection chemical has been largely in the hands of the chemist. The aim appears to be to find a molecule with activity, modify its structure in as many ways as possible until the maximum activity is found by bioassay and then develop the promising chemicals further. Few companies have considered natural chemicals as control agents, e.g. little attempt has been made to harness the fungistatic activity of ammonia (section 7.2) or the vast range of antibiotics produced by micro-organisms. It is perhaps ironic that the pharmaceutical industry has paid more attention to soil micro-organisms, particularly actinomycetes, for its source of chemicals than has the agrochemical industry. One exception is the gibberellin story (section 7.2) but even then many plant physiologists appear to have forgotten the original isolation from micro-organisms.

A simple chemical with some promise is calcium peroxide, which can be manufactured fairly cheaply from lime and is obtained as a formulation with lime [40]. A greenhouse trial with the formulation is shown in Fig. 10.1. When applied as a seed coating it breaks down in the presence of soil water to yield oxygen very slowly over a period of weeks. Where the seed is stressed by low concentrations of available oxygen, perhaps because of microbial competition, the dressing can provide the oxygen needed for germination. This has been particularly valuable for rice which has not been pre-germinated [52]. As the material breaks down, alkali is formed and this can neutralize organic acid toxins. In addition, the material itself is antifungal, probably because most micro-organisms do not have sufficient

Fig. 10.1 Effect of coating wheat seed with a formulation containing calcium peroxide on the emergence from wet soil with straw residues. Seeds which were coated with the formulation (C) emerged, whereas those which were uncoated (O) or coated with the polyvinyl alcohol binder only (P) failed to emerge. (Photograph by M. Sladdin, Letcombe Laboratory.)

catalase to break down the hydrogen peroxide formed as an intermediate in breakdown. This action could be a problem with legumes inoculated with *Rhizobium* because hydrogen peroxide is also toxic to the bacterium. There would be few environmental problems with such a chemical. Any chemical is effectively applied as an insurance and is only active if the particular stress occurs in that year. Therefore, the farmer and agricultural adviser must assess the risks of crop damage against the cost of that insurance.

10.2   **Control of *Fusarium* foot-rot of wheat in a wheat–fallow cycle**

Traditionally, soil-borne plant pathogens were restrained in soils by cultivation practices such as a long (about four years) sequence of crop rotations. However, it is now possible to minimize *Fusarium* foot-rot of wheat in the Pacific North-west of the USA by a two-year rotation which differs little from the existing method of farming (Fig. 10.2) [15].

The disease is caused by *Fusarium culmorum* (*F. roseum* 'culmorum') or *F. graminearum* (*F. roseum* 'Graminearum') with chlamydospores as the primary inoculum on plant debris in the mulched surface layer (*c.* 10 cm deep). The disease occurs on dry-land wheat grains, grown on the fallow. The dry soils in the autumn are ideal for

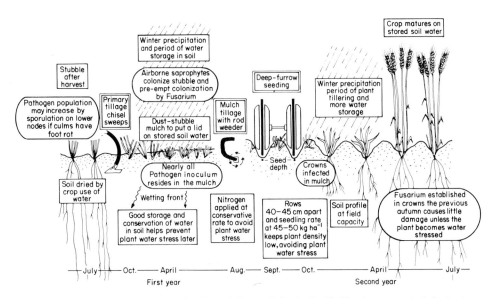

Fig. 10.2 How the wheat–fallow cycle in the Pacific North-west controls foot-rot caused by *Fusarium roseum* 'Culmorum'. (From Cook [15].)

infection because the fungus functions optimally at low soil-water potentials [53]. The brown colour of the diseased crown can be seen by cutting it longitudinally.

In the wheat–fallow management, a field harvested in the summer of one year is not planted again until the summer of the following year; therefore, the crop is produced on the net accumulation of precipitation for two years. The disease is associated with plant-water stress because; (1) it is greater on slopes compared with the flat; (2) the blight does not occur even with a high incidence of infection if there is much rainfall in the late spring; and (3) plants growing with fertilizer nitrogen are more diseased than those growing on residual soil nitrogen only. The latter observation suggested that the key to the control of the disease might be by management of the rate of nitrogen application. The plant gets a lower midday leaf-water potential and is therefore stressed when application rates are high because the leaf area is greater. By increasing row spacing of the crop, the loss of water by transpiration per unit area of field is decreased, reducing the plant-water stress and the disease.

The system for control which has been devised is specific to *Fusarium* foot-rot on dry-land wheat in the Pacific North-west and operates well in practice. It cannot be expected to control the disease in other areas or other kinds of *Fusarium* damage but it illustrates well the type of integrated approach needed by plant and soil scientists which may be effective in other problems.

10.3   **Biological control by soil transfer**

Papaya is grown on lava-rock land on the island of Hawaii. When the trees become tall and difficult to manage and yields decline (about three to four years after planting), replanting is usually not possible because the new plants would become infected with a root-rot caused by *Phytophthora palmivora*. Originally, although there was an extensive acreage of abandoned papaya fields, there was also enough virgin land for replanting. When the virgin land became exhausted, alternative solutions had to be found [39].

The infection of papaya by *P. palmivora* occurs on fruits and the upper portion of the trunk during rainy periods. The diseased fruits, with a cover of sporangia and chlamydospores, fall to the ground and act as the inoculum source in the soil when replanting is attempted. No resistant cultivar is available, the rocky and porous nature of the land make it inappropriate to fumigate and fungicides applied as soil drenches are unsuccessful.

In field tests, it was shown [39] that when papayas were planted in holes (30 cm in diameter and 10 cm deep) filled with virgin soil and surrounded by soil previously planted with papaya, none of the plants became infected. In contrast, an average of 33 per cent of those planted in the surrounding soil became infected (Fig. 10.3). One year after planting, all the papayas in the 'virgin islands' were healthy and produced fruits.

Fig. 10.3 Control of papaya root-rot caused by *Phytophthora palmivora*.
VS = papaya trees grown in a small quantity of virgin soil placed in planting holes in the replant field; C = the control plants grown in planting holes without virgin soil. The plants are one year old. (Photograph by W-H. Ko, University of Hawaii.)

The observations suggested that mature roots of the plant are resistant to infection and this was confirmed in a greenhouse study. It was suggested that the virgin soil is fungistatic to the many mycelium or spores attempting to enter it during the plant establishment. This method of biological control by soil transfer is now used extensively in Hawaii and the principle involved might be applied to other crops.

Similar observations on soil transfer have been made with take-all (section 6.4) [59]. Soil taken from fields in which the disease does not occur ('suppressive soil') can eliminate the disease when added to soil where the disease is rampant ('conducive soil'). The mechanism suggested here is that bacterial antagonists of the fungal pathogen are transferred in the suppressive soil. Whereas this would be quite impracticable as a biological control procedure for fields of wheat, it has led to the idea of biological control by seed inoculation or soil

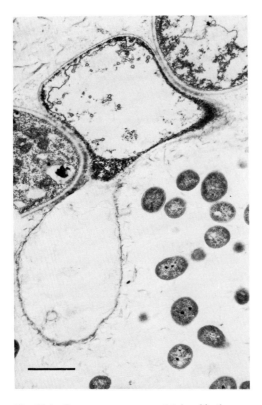

Fig. 10.4 *Gaeumannomyces graminis* lysed in the presence of a *Pseudomonas* sp. The culture is from a sand–montmorillonite mixture. The montmorillonite is adsorbed on to the fungus but not on to the bacterium. Bar marker = 1 μm. (Photograph by R. Campbell, Bristol University.)

drenches (section 10.4). Indirectly, a similar effect might be achieved by crop rotation [16]. When beans are introduced following wheat, soil is not only made conducive but a large inoculum of the pathogen is maintained in soil. Other crops can induce various degrees of suppressiveness. It is assumed, but not demonstrated, that the soil population changes under different crops but it is not proven whether the rhizosphere micro-organisms or the microflora of decomposing plant residues is responsible for the decline in the disease. Various factors were suggested as potential explanations of take-all decline [17]. Hypovirulence (the loss of virulence of the pathogen) occurs in laboratory culture but there is little evidence for this in the field. Cross-protection by an avirulent fungus, *Phialophora graminicola*, is possible but no evidence was found for this in the Palouse soils. Predation by amoeba of the family Vampyrellidae is restricted to specific water potentials but good evidence was provided for protection by fluorescent pseudomonads (Fig. 10.4). Other evidence [28] shows that the control can be achieved with *Bacillus mycoides* (Fig. 10.5).

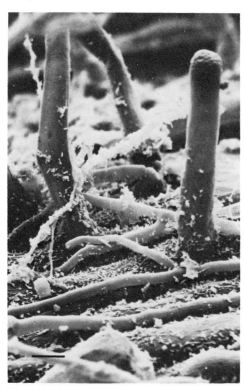

Fig. 10.5 *Gaeumannomyces graminis* on a wheat root in sand culture inoculated with *Bacillus mycoides*. Bar marker = 10 μm. (Photograph by R. Campbell, Bristol University.)

10.4   **Inoculation**

The soil biomass can be modified by varying agricultural practices (section 9.1). To modify the size of the biomass by inoculation would be a much more difficult prospect. If the biomass in the surface 5 cm of an average soil is about 300 kg C ha^{-1}, then about 6 t fresh weight of bacteria ha^{-1} would need to be added to double the biomass. This is about equivalent to the weight of straw normally returned to the soil and as biomass yields from substrates are no greater than 40 per cent it is unworkable. Feasibility depends on the function required of the inoculated biomass. If the action required is a general one, such as the stabilization of soil aggregates (section 2.2), then inoculation is unlikely to be practical unless the organisms added are dramatically good at producing aggregates and are good competitors against the other components of the soil microflora. The prospects would be more hopeful if the action were more subtle, e.g. if the inoculated organisms were antagonistic to pathogens, mimicking the soil transfer process (section 10.3). Even then, however, there is probably much greater promise if the organisms are taken to the site of the action, e.g. if the problem is in seedling establishment, it would probably be more effective to inoculate seed with organisms which will colonize the young seedling roots. The principles of the various types of biological control are discussed in detail by Baker and Cook [3] and a second edition of this source book will appear shortly. The strategy for the management of associated biota as a means of biological control [14] is: (1) reduce the inoculum density of the pathogen or prevent it establishing, which usually requires some predisposing stress such as flooding; (2) replace the pathogen in plant-refuse substrate by allowing a saprophyte to colonize, e.g. this can sometimes be achieved by the modification of the water potential; (3) suppress germination or growth of the pathogen; (4) protect the infection court, e.g. with mycorrhiza; and (5) stimulate a resistance response in a potential host.

10.4.1   *Seed bacterization by Azotobacter and phosphate solubilizers*

The idea of increasing the nitrogen-fixing populations in soil has always appealed to microbiologists and soil scientists. Similarly, in some parts of the world where phosphorus is present but unavailable to plants, an increase in the phosphate-solubilizing population would be of great value.

The early attempts at increasing nitrogen fixation were made in the USSR by inoculating *Azotobacter chroococcum* ('azotobacterin') on cereal seeds [49]. After some promising early results, the general pattern of results showed that the treatment effects were evenly divided between positive, negative and no effect. The same pattern emerged in Britain [8], Australia [57,62] and India [47]. Several investigators have attempted to analyse the reasons for these variations but without much success. Certainly there seems little prospect for the dinitrogen fixation contributing much to the stimulations in plant growth. The more recent work on the inoculation of tropical grasses with associative nitrogen fixers, either *Azotobacter paspali* on to roots of the sand grass *Paspalum notatum* [22] or *Azospirillum lipoferum* on to *Digitaria decumbens* [21], may be of some significance but it remains to be proven that any response is due to dinitrogen fixation. The proposal that the tropical grasses using the C_4 photosynthetic pathway would release more carbon than those using the conventional C_3 pathway, thereby providing more energy for fixation, has not yet been substantiated. A major difficulty is that so many of the studies have used the acetylene-reduction test as the index of dinitrogen fixation and there are a large number of problems in the interpretation of this test. Incorporation of $^{15}N_2$ into the plant cells or plant response are the only positive indexes of fixation but the increase in nitrogen by chemical assay is also a useful index. Some of the most useful experiments have been undertaken by Dr G. Lethbridge of the Macaulay Institute for Soil Research (personal communication). He used ^{15}N dilution to estimate dinitrogen fixation in glasshouse-grown spring wheat inoculated with dinitrogen-fixing bacteria (*Azotobacter, Azospirillum, Klebsiella* and *Bacillus* spp.) isolated from the roots of wheat or other grasses in Scotland, Brazil or the USA. Experiments were done under gnotobiotic and non-sterile conditions, at nitrogen levels ranging from 1–56 mg nitrogen per plant as Ca $(^{15}NO_3)_2$ in sand culture and three soils of differing nitrogen contents. In sand culture, 99 per cent ^{15}N-labelled wheat seeds were also used and if the inoculum was large this too was labelled, to ensure the only ^{14}N source was atmospheric dinitrogen. This was necessary because some of the experiments were done under conditions of severe nitrogen limitation, the most commonly reported conditions under which inoculation effects are claimed by others. In these circumstances the nitrogen reserve of the seed represented up to 30 per cent of the final nitrogen content of the plant and it did not prove possible to accurately predict seed nitrogen content from its weight. Plants were grown to maturity and roots, straw and grain were analysed separately. Root-associated dinitrogen fixation was negligible unless a carbon source was added to the rooting medium. When large amounts of microbial biomass were added to the roots, the roots used this as a source of nitrogen. These observations demonstrate that despite the potential of rhizosphere bacteria to fix a significant amount of nitrogen which the plant can use, the plant appears unable to supply enough carbohydrate to make the system function. The best prospect for the associative dinitrogen fixation being of significance is probably when plants fight for survival in the ecosystem and where nutrients are sparse, such as in the sand-dune grasses [1].

Brown [6] has argued that the major responses to inoculation with dinitrogen fixers are not due to additional nitrogen, but to the provision of growth regulators, particularly gibberellic acid [7] (section 7.2). Some of the growth responses are indicative of the growth regulator pools in the plant being modified. However, roots of cereals, compared with the tomatoes used in some of Brown's studies, seem rather insensitive to growth regulators when they are applied exogenously [45].

A problem with most attempts at seed inoculation is that the physiology and growth of the inoculum is not controlled. When *Azotobacter chroococcum* was grown in a chemostat, cultures which had been grown with a source of fixed nitrogen systematically inhibited growth, probably by oxygen competition, but those which fixed nitrogen sometimes provided the stimulation in growth [30]. Therefore, the mechanisms and control of the stimulation of plant growth by *Azotobacter* inoculation remain elusive.

Some bacteria, particularly those which produce 2-keto-gluconic acid, have the potential to solubilize phosphorus [23]. The preparation 'phosphobacterin' was composed of *Bacillus megatherium* var. *phosphaticum* [18] which is usually present in soil as spores and is therefore a bad choice of inoculant, being normally non-functional in the rhizosphere. Furthermore, it is less effective in phosphorus solubilization than most rhizosphere bacteria and was generally added to seeds at very low densities (about 4×10^4 cells seed^{-1}). Needless to say, little value was found from such a poorly conceived concept.

10.4.2   *Enhanced plant growth by siderophores from plant-growth-promoting rhizobacteria (PGPR)*

In some of the early studies on seed inoculation, it was recognized that the action of the inoculant might be to prevent the colonization of the root and seed by pathogens or other antagonists but there were few reports to substantiate this. However, it has now been shown that specific strains of the *Pseudomonas fluorescens-putida* group rapidly colonize plant roots of potato, sugar-beet and radish and cause statistically significant yield increases (up to 144 per cent) in field tests [36, 37, 38]. The PGPR promote plant growth indirectly by depriving the native microflora of iron, so making it less available to certain members of the natural microflora, particularly pathogens. The agents in chelating the iron are extracellular siderophores, which are microbial iron transport agents. Normally, these compounds occur in a wide range of soils and their concentration·is correlated with per cent organic matter [55]. They probably solubilize ferric oxides to make the iron available to plant roots. However, it seems that in the rhizosphere, if the extracellular products of the bacteria can chelate the iron and then subsequently make it available to plants, avoiding the pathogen or subclinical pathogen, then plant growth will benefit.

The new siderophore produced by the PGPR is pseudobactin and this was shown in *in vitro* experiments to inhibit the growth of the plant pathogen *Erwina carotovora*, which produces soft-rot of potato and seedpiece decay [36]. This did not occur if the growth medium was amended with ferric chloride. Grown *in vivo* in the greenhouse, with an addition of $10 \, \mathrm{m \, mol \, m^{-3}}$ pseudobactin to the water supply, led to more than a doubling of the weight of potato plants (Table

Table 10.1 Promotion of potato growth by the *Pseudomonas* strain B10 and a pseudobactin in a greenhouse assay. (After Kloepper *et al.* [36].)

Treatment	Average plant weight (g)
Water control	1·0
$50 \, \mathrm{m \, mol \, m^{-3}}$ Fe^{III} EDTA	0·9
Pseudomonas	2·3*
Pseudomonas $+ 50 \, \mathrm{m \, mol \, m^{-3}}$ Fe^{III} EDTA	0·9
$10 \, \mathrm{m \, mol \, m^{-3}}$ pseudobactin	2·5*
$50 \, \mathrm{m \, mol \, m^{-3}}$ ferric pseudobactin	1·1

* Significantly different from the control ($P = 0.01$).

10.1) and this was attributed to a 74 per cent decrease in the number of pathogenic fungi found in the root region. The PGPR did not exhibit antibiosis against *Escherichia coli* K-12 AN 193 which also produces a siderophore, enterobactin. However, it was active against the parent K12-194, which does not produce the siderophore. Mutants of the PGPR which are unable to produce the pseudobactin are also inactive against *E. carotovora*.

A problem with the interpretation of inoculant effects is whether the introduced bacterium colonizes the entire root system. In the studies on potato, the two *Pseudomonas* spp. inoculated were marked genetically by selecting strains with resistance to the antibiotics rifampicin and nalidixic acid [37]. By this means the bacteria could be isolated selectively from the roots of the growing potato plant. This showed that the bacteria colonized the entire rhizosphere of treated plants, including developing daughter tubers and the apical roots of adjacent untreated plants. The populations of the inoculated bacteria were as great as 9.6×10^5 cm^{-1} root two weeks after plant emergence and averaged 10^3 cm^{-1} throughout the season, but declined to 10^2 cm^{-1} at harvest. These counts were relatively constant throughout the root system. The inoculants increased the growth of the potato plant by up to 500 per cent in the greenhouse and significant yield increases (up to 17 per cent) were obtained in four of five field trials. The increase in the early growth in the field is evident from Fig. 10.6.

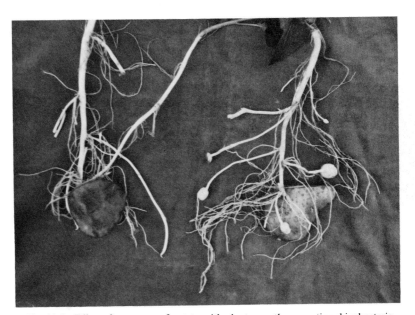

Fig. 10.6 Effect of treatment of potato with plant-growth-promoting rhizobacteria (right) compared with untreated plants (left). Note the extra development of tubers in response to the inoculation. (Photograph by J. W. Kloepper, University of California.)

It is generally considered that the rhizosphere of each plant species is distinctive and therefore it would be somewhat surprising if the potato inoculants colonized roots of other species. Even so, it was shown that when the fluorescent *Pseudomonas* strain B10 (which was isolated from potato growing in take-all suppressive soil) and its associated siderophore were added to soil they were both effective in preventing disease of barley in take-all conducive soil inoculated within *Gaeumannomyces graminis* var. tritici and in *Fusarium*-wilt conducive soil inoculated with *Fusarium oxysporum* f.sp. *lini*. When exogenous iron (III) was added to the disease-suppressive soil, it again became conducive, probably because siderophore production was repressed. The generally useful action of the siderophores in suppressing disease can be seen to apply fairly widely.

The effect of siderophores may be complicated in anaerobic soils (section 2.3). In these, the formation of organic acids causes iron to be solubilized and it also triggers the formation of ethylene. The former action might repress siderophore production and make roots more susceptible to disease. Take-all and some other diseases are certainly more prevalent in wet soils whereas *Fusarium* disease is more common in dry soils because of the water-potential optimum (section 10.2). The stimulation of ethylene formation in these soils could reduce root growth (section 7.2).

10.4.3   *Biological control of crown gall by agrocin 84*

Crown gall, besides affecting cruciferae also affects fruit trees; almond, peach and rose are the most susceptible. In studying strains of *Agrobacterium radiobacter* var. *tumefaciens* responsible for the disease, it was found that some strains were non-pathogenic [35]. One strain, 84, is now used in biological control, its effectiveness depending on achieving a ratio of pathogen to non-pathogen of less than one. The organism is supplied on a peat base, similar to *Rhizobium* inoculants (section 6.2), or on agar and can be used on seeds and roots.

The non-pathogen produces an antibiotic, agrocin 84, being a fraudulent adenine nucleotide with two substitutions, although the complete chemical structure has not yet been assigned:

The production is controlled by a large plasmid.

Whereas disease control has been very satisfactory in Australia, some *Agrobacterium* strains isolated from peach in Greece cannot be controlled with agrocin 84. The reason is that the genes controlling agrocin 84 production (and resistance) can be transferred to a pathogenic recipient but only in the presence of nopaline, which is contained in crown-gall tissue. The effectiveness of the control in Australia appears to be because the initial effectiveness has eliminated all the gall tissue, whereas in Greece artificial inoculation with pathogens provided a situation where strain 84 did not give total control, galls developed, nopaline was produced and plasmid transfer occurred. It appears that a mutant of strain 84 with a defective plasmid transfer or mobilization system will have to be sought. Such approaches will continue to make exciting science and to maintain confidence in the commercial application of biological control.

10.5   **Microbial insecticides**

Despite the commercial success of *Bacillus thurigiensis* toxin and *Verticillium lecanii* against insects colonizing leaves, there have been few attempts to produce such an agent against soil-inhabiting insect pests. *Beauvaria bassiana* is an entomopathogenic fungus considered to be potentially effective for the control of soil-inhabiting insect pests but the problem in its exploitation appears to be that the conidia only survive in soil for short periods. It has been suggested [58] that patulin, produced by the fungus *Penicillium urticae*, is the factor which inhibits the germination of *B. bassiana* in soil. However, as with earlier claims for patulin as a phytotoxin in soil (section 7.3), the true ecological significance of this has yet to be evaluated because the relationships between concentrations found in soil and biological activity have yet to be fully evaluated.

10.6   **Nematode-trapping fungi**

It has long been recognized that certain fungi have the ability to trap nematodes, either by the powerful adhesives which they produce or by producing rings or wedges around the prey (Fig. 10.7). However, there have been few attempts at the commercial exploitation of these processes in biological control measures. The fungus *Arthrobotyrs conoides* has been shown to have some activity in untreated soil [27] but it was particularly active in reducing nematode numbers after two weeks in soil supplemented with alfalfa meal. Even then, the nematodes returned to their original numbers after four weeks. Obviously, there is potential for control but exploitation would require some cunning to maintain the activity.

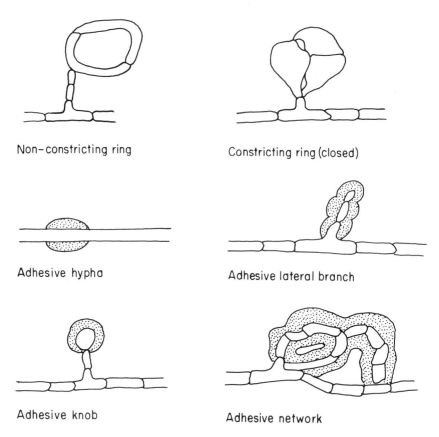

Non–constricting ring   Constricting ring (closed)

Adhesive hypha   Adhesive lateral branch

Adhesive knob   Adhesive network

Fig. 10.7   Mechanisms for the trapping of nematodes by fungi. Shaded areas = adhesive materials.

10.7   **Composting and organic farming**

The use of chemicals in agriculture is now regarded as conventional but previously the farmer had no option but to use methods which we now term organic farming. The true organic farming movement, as we understand it today, originated with Sir Albert Howard [32], a British plant scientist who developed the Indore process in India for the manufacture of compost from vegetable and animal wastes. Unfortunately, he isolated himself from the exciting scientific developments taking place in world agriculture and his book was both emotive and inaccurate, as is disclosed by the following: 'The Industrial Revolution, by creating a new hunger—that of the machine— and a vast increase in the urban population, has encroached seriously on the world's store of fertility. A rapid transfer of the soil's capital is taking place. The soils of the world are either being worn out and left in ruins, or are being slowly poisoned.' Emotive and unreasoned arguments have since become the hallmark of many in the organic farming movement. It has, therefore, become very easy for distinguished soil scientists such as F. E. Allison [2] to provide ammunition for its destruction, by pointing to the lack of scientific evidence in its favour. However, in showing the futility of the arguments of Howard and others, it is easy to ignore the possibility that within the organic farming movement there may be some useful lessons which could help to improve modern agriculture. Allison, for example, cites Balfour [4] as stating, 'soil biologists have tended to focus their attention on specialized groups of soil organisms such as, for example, nitrogen-fixing bacteria, and specific parasitic fungi. They have largely overlooked the general relationship of the soil microflora with normal growth'. Allison takes this as a criticism of the excellent pure culture studies in the laboratory with nitrogen fixers and pathogens, and to an extent his concern is fair. However, Lady Balfour also has a point and, indeed, much can be gained from studying the effects of mixed populations of non-symbiotic micro-organisms on plant growth, such as their role in nutrient uptake or root exudation (Chapter 5).

Scientific assessments of organic farming have been few but two important study groups in the USA with teams of distinguished scientists, one sponsored by the United States Department of Agriculture (USDA) [63] and the other by the Council for Agriculture Science and Technology (CAST) [19], published their findings in 1980. In their comparison of conventional and organic farming methods (Table 10.2), they reached different conclusions. The CAST report made an either/or assessment and demonstrated that it would be economic suicide to change to organic methods on a large scale.

Table 10.2   Comparison of conventional and organic farming.

Conventional	Organic
Use of commercial fertilizers and animal-feed additives	Use of natural resources, particularly manures, but also phosphate rock and limestone
Monoculture common	Mixed farming and crop rotation common
Few legumes	Extensive use of leguminous crops to supply nitrogen
More fossil fuel used per hectare, mainly because chemicals used	Similar energy use 'on-farm'
Use of synthetic herbicides, pesticides and growth regulators	Biological weed and pest control
Larger financial inputs, greater profits	Smaller inputs, less return

The USDA report was far more open and recognized that there are problems with conventional methods and the introduction of some of the organic ideas might be profitable. They also recognized that not all organic farmers were against the use of chemicals but opt for a minimum input system to get maximum returns. We should accept that some of the conventional systems are not ideal and that there is scope for optimization by the introduction of new or old technologies.

A point of disagreement in the two reports centres on soil organic matter and soil structure. The USDA report provided evidence (Table 10.3) that corn monoculture decreased the soil organic-matter

Table 10.3 Soil organic matter as influenced by the cropping sequence. (After Smith [60].)

Cropping sequence	% Soil organic matter content (no manure applied)*
Continuous	
Corn	1·45
Wheat	3·40
Oats	4.08
Timothy	4·68
Rotation	
3 year corn/wheat/red clover	3·31
4 year corn/oats/wheat/red clover	3·74
6 year corn/oats/wheat/red clover/timothy/ timothy	3·83
Virgin	
Mixed grass and timber	5·78

* Original data as % nitrogen converted to % organic matter by multiplying by 17.

content compared with other crops or with rotations including corn, whereas monoculture with timothy had least effect on the organic matter. In some organic systems, with an input of carbon in the form of animal wastes and manures, the microbial degradation of the extra substrate probably yields more aggregating agents and hence improves soil structure. In contrast, the CAST report indicated that soil erosion would increase if organic methods were widely adopted because legumes would have to be grown on level land; the extra row-crops would have to be planted on slopes where erosion potential is greater.

Existing conventional systems are governed by economics and on this basis there are usually few criticisms at present. However, economics change, e.g. the production of inorganic fertilizers is energy intensive and subject to market fluctuations which have increased greatly in recent years. We must be prepared for change if the need arises. Organic systems include the composting of plant and animal wastes, the science of which is generally poorly understood. There has been little incentive to acquire knowledge on these complex processes because in many conventional systems, such as in Britain, monoculture of crops and animal production are separated by great distances. It is uneconomical to transport the animal waste to use as compost or to transport the plant waste to upgrade as fodder. A recent symposium report [61] identified a need for the scientific approach to biological husbandry (organic farming). Any introduction of aspects of organic methods to conventional systems should not be seen as a threat to productivity in the agrochemical industry but as complementary, to attain even greater yields.

Measurements of microbiological activity in relation to organic farming have been few but some preliminary observations have been reported (Table 10.4).

Table 10.4 Comparison of biochemical and microbiological properties of soil taken from adjacent farms on a Naft silt loam in Washington state, USA. Winter wheat/spring pea was the normal rotation on both farms. Inorganic fertilizers have not been used on the organic farm since 1909 and plant nutrients are derived from a green manure of Austrian winter peas during a summer fallow; plant protection chemicals are used on this farm. Yields on the two farms are similar. (Data from H. Bolton Jr, MS thesis, Washington State University.)

	Organic farm	Conventional farm
Nitrogen mineralization		
$NH_4-N+NO_3-N \mu g\, g^{-1}$	39·2	43·5
Phosphatase μg p-nitrophenol g^{-1}	32700·0	26600·0
Dehydrogenase μg triphenylformazan g^{-1}	29·3*	21·7
Urease $\mu g\, NH_4-N\, g^{-1}$	30·4*	21·7
Biomass mg C $100\, g^{-1}$	15·25	9·4
Nitrogen mineralization after $CHCl_3$ fumigation $NH_4-N+NO_3-N \mu g\, g^{-1}$	14·5*	6·7
Plate counts g^{-1}		
Fungi $\times 10^5$	3·0	2·4
Bacteria $\times 10^7$	2·1	1·8
Actinomycetes $\times 10^6$	3·3	2·8
Denitrifiers MPN $\times 10^5$	2·8	1·9
Nitrifiers MPN		
Nitrosomonas spp. $\times 10^4$	11·3	79·2*
Nitrobacter spp. $\times 10^2$	4·6*	12·2
Organic carbon % w/w	0·95	0·91
Total nitrogen $\mu g\, g^{-1}$	1050·0	1037·0
Soil pH	5·97	5·77

* Significant differences ($P \leqslant 0.05$).

What is the scope for controlled microbiological breakdown of plant wastes in low-technology on-farm processes? There appear to be at least three areas for profitable research.

10.7.1   *Production of soil-stabilizing agents*

The idea that micro-organisms are major agents in promoting soil structure was discussed in section 2.2. The soil biomass increases when straw is added to soil or if a grass ley is introduced [43]. The extra biomass probably improves soil structure but it may be possible to produce an inoculant compost with a high population of polysaccharide producers which would produce more aggregating agents per unit weight of available substrate. Micro-organisms degrading straw in nutrient media [41] cemented and bound aggregates (Fig. 10.8). This increased the stabilization of an unstable silt loam soil and produced structure in volcanic ash from Mount St Helens where little or none had existed previously (Table 10.5). Primordial soils in the Pacific North-west originated from volcanic ash and glacial morraine and their structure was achieved after the first plants were grown and micro-organisms degraded their residues. This is a continuing process but a dynamic balance because the aggregating agents can be destroyed by microbial action [46]. It is therefore essential that we keep the balance in favour of the producers.

Table 10.5   Water stability and water retention of ash and soil aggregates. (From Lynch & Elliott [41].)

Treatment	% solid in suspension*		Water retention (g g^{-1} solid)*	
	Soil	Ash	Soil	Ash
Degraded straw	8·5a	25·0c	0·379w	0·461x
Undegraded straw	11·0b	38·2d	0·370w	0·437y
None (control)	13·6b	40·5d	0·378w	0·449z

* Figures not followed by the same letter are significantly different ($P=0.05$) using Duncan's multiple range test for a variable sample.

10.7.2   *Upgrading of the nitrogen value of residues*

It was indicated in section 6.2 that the energy (sugar) requirement for dinitrogen fixation is great and that the provision of sugar by roots is unlikely to make a major contribution to agricultural needs.

Fig. 10.8   Soil aggregation by micro-organisms. (a) Fungal hypha with adhering soil. (b) Bacteria on the surface of straw with adhering ash. (c) Microbial gum with adhering soil. Bar markers = 10 μm.

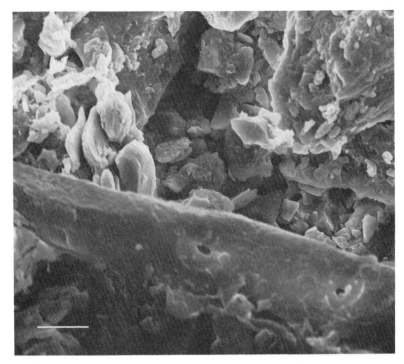

10.8a

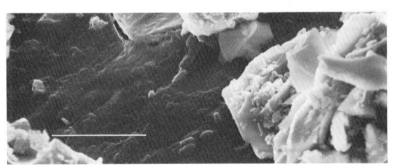

10.8b

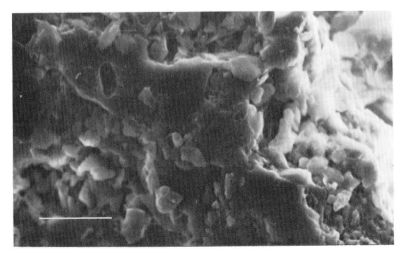

10.8c

Straw, by comparison, has an abundant source of sugar as cellulose and hemicellulose, which is, in effect, wasted. If this could be channelled to dinitrogen fixation by an on-farm composting process, the dinitrogen contained in the microbial source might make a valuable contribution to fertilizer needs.

In systems in the laboratory where no attempts have been made to optimize the process by inoculation or control of the growth environment, the potential is difficult to estimate because: (a) studies are often limited to the early stages of straw decomposition and weight losses are not recorded; (b) methodology is sometimes doubtful, using only the acetylene-reduction test; and (c) studies with pure cellulose as the carbon source do not necessarily relate directly to straw, e.g. the gain of 2–3 mg N g^{-1} straw (equivalent to 2–3 kg t^{-1}) reported by Rice and Paul [56] is actually 13–16 mg N g^{-1} straw used when allowance is made for the incomplete degradation. In fact, it seems reasonable that 50 per cent of the straw could be degraded in a compost (section 4.2) and that this would give about 6 mg N g^{-1} total straw. It is difficult to guess by how much would optimization increase efficiency but 15 mg N g^{-1} total straw would seem a reasonable target. When this amount is added to the amount already present in the straw (c. 5 kg N t^{-1}) the total nitrogen value could become 20 kg N t^{-1}. In Britain, arable crops produce about 7 t ha^{-1} and therefore the 140 kg nitrogen would virtually satisfy fertilizer needs. The analysis can also be made at the cell level by assuming a carbon/

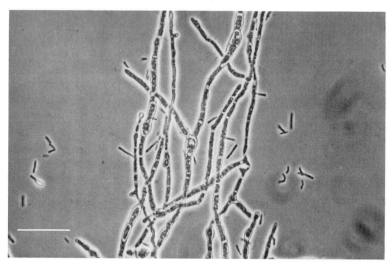

Fig. 10.9 Growth of the anaerobe *Clostridium butyricum* in association with *Trichoderma viride* in the presence of air. Bar marker = 1 μm. (Photograph by D. A. Veal, Letcombe Laboratory.)

nitrogen ratio of the microbial cells as $5:1$, a yield factor of 0.35 and an optional fixation rate of $36\,mg\ N$ fixed g^{-1} substrate consumed [54]. This again shows that $15\,mg\ N\ g^{-1}$ total straw would be fixed.

A basic problem is that *Azotobacter* does not have a cellulase, so that the fermentation would either have to be linked with a cellulase-positive organism, such as *Trichoderma*, to yield the cellobiose units or, alternatively, a cellulolytic nitrogen-fixing anaerobe, such as *Clostridium*, would have to be found (Fig. 10.9). The latter might be easiest to achieve in an 'anaerobic' ditch. Even in an aerobic process, aeration must be minimized to achieve the respiratory protection to the nitrogenase (section 6.2). The efficiency of dinitrogen fixation by mixed cultures of *Azotobacter*, *Clostridium* and *Cellovibrio* (a cellulolytic aerobe without a nitrogenase) is similar to that of *Azotobacter* in pure culture [34] and the mixed culture fermentation probably has the greatest prospects.

10.7.3   *Minimization of pathogen and phytotoxicity problems*

Plant residues serve as a substrate for organisms in the soil which are both antagonistic and beneficial to plant growth. The aim of a compost must be to switch the balance towards the latter. When the depleted substrate (compost) is then introduced to the soil, there is less potential for the pathogens and phytotoxin producers to grow. It is well recognized that many plant pathogens survive on plant residues but the usual approach is to bury them by ploughing. The phytotoxin producers are particularly active during the early stages of straw decomposition [12,31]. In simple laboratory tests it has been shown that by degrading the straw, the toxin-producing potential is minimized [42]. This effect is not sensitive to inoculation of any particular microbial group and is even more effective if lime is added to the straw. Composting in windrows in the field would possibly be an approach to achieve this. There is little experimental evidence that pathogens would also be eliminated in this way. However, it is interesting that organic farmers using composts as a basis for their system seldom encounter major pathogen problems and this warrants further investigation. *Trichoderma harzianum* is a particularly strong colonizer of straw because, like other species of *Trichoderma* [44], it exhibits a powerful cellulase activity. It is, therefore, of interest that this fungus is effective in the control of a variety of root pathogens [13].

10.8   **Useful properties of algae for agriculture**

The value of algae in agriculture has been reviewed by Metting [48]. Inevitably the blue-green algae can make a nitrogen input to soils if there is an abundant supply of water and sunlight on the soil surface. Rice soils have benefited greatly from algal inoculation and in countries where nitrogen fertilizers are at a premium, this biological source is critical. As yet, yields with algal inocula are seldom as great as with fertilizer. In the long term the value of inocula may depend on the rate at which the algal nitrogen becomes available for plant uptake. In temperate zones, little immediate input of nitrogen by algal inoculation has been observed. However, of particular potential for exploitation is the symbiotic association of the freshwater fern *Azolla* with the blue-green alga *Anabaena azollae*, which has been reported to produce between 100 and 700 kg N ha^{-1} year^{-1}. Provided all this is available to the crop it would meet the fertilizer need [10].

Blue-green and green algae have been used as green manure in Tashkent, for reclamation of saline and sodic-saline soils in India and to improve soil structure in Central and North America. The algae can increase the carbon dioxide in the soil atmosphere, decreasing the pH and solubilizing calcium (as the carbonate), which replaces sodium adsorbed to clays and therefore lessens soil dispersion. Algae can also produce large amounts of extracellular polymers which have potential value as aggregating agents (Fig. 10.10). Even though chemical soil conditioners are generally uneconomic except for specific horticultural crops, the algal source using sunlight as an energy source might be cheaper, especially if the nitrogen value is deducted from the cost. However, even greater potential exists if the algal populations increase after they have been added. This again depends on the presence of light and moisture, such as commonly occurs in irrigated agricultural systems where soil conditioners would be particularly valuable.

Fig. 10.10 Scanning electron micrograph of an algal crust on the surface of undisturbed soil. u = unicellular alga; b = blue-green alga; d = diatom. The specimens in a and b were prepared by cryopreservation. The crust is composed mainly of a unicellular alga embedded in a hydrated matrix and overlaid by a distinct layer (a). The matrix and overlayer are thought to be polysaccharide. Dehydration of the specimen on the heated stage of the SEM (b) causes the overlayer to peel away and the matrix to shrink, revealing more detail of the cells in the crust. Severe dehydration of a similar specimen by critical point drying (c) causes the overlayer and matrix to disappear almost totally leaving only residues, seen as small aggregates and strands on the cell surfaces. Bar markers = (a) 30 μm; (b) 100 μm; (c) 50 μm. (Photographs of a and b by J. A. Sargent and of c by J. Bebb, ARC Weed Research Organization, Oxford.)

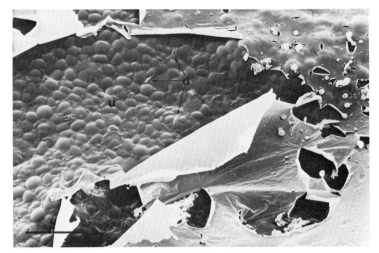

10.10a

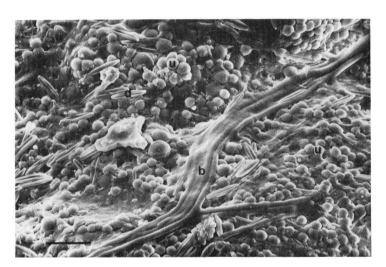

10.10b

10.10c

10.9   **Genetic modification of the plant**

One of the major approaches to plant breeding and the consequent increased yields that have been obtained in the so-called 'green revolution' is that cultivars have been selected which are resistant to the major diseases [5]. This remains one of the foremost considerations in the selection of a new genotype. However, it has seldom been considered that a plant genotype could influence other microbial activities.

The substitution of a pair of 5B chromosomes into a wheat cultivar changed the rhizosphere characteristics (root rotting, cellulolytic, pectinolytic, amylolytic, ammonifying and total bacteria) so that the recipient plant had a rhizosphere similar to that of the donor parent *Azotobacter paspali* [51] established on the rhizoplane of tetraploid but not on that of the diploid cultivars of the tropical grass *Paspalum notatum* [20]. Rice cultivars introduced before the use of inorganic nitrogen fertilizer possess greater nitrogenase activity than those in current use [29]. The old varieties of wheat used in the Pacific North-west appear more resistant to the effects of subclinical pathogens (pseudomonads) than the more recent varieties (Table 10.6) [26]. All these effects could be a consequence of the different forms

Table 10.6   Effect of two inhibitory *Pseudomonas* sp. on cultivars of winter-wheat seedling growth. Each result represents the mean of ten bioassays. (After Elliott & Lynch [26].)

Wheat cultivar	Approximate era in Pacific North-west	*Pseudomonas* sp. (NT-20)	*Pseudomonas* sp. (NT-13)	Control root length (mm)
		(% control)		
Jones Fife	1919	88	98	51·8
Little Club	1919	87	114	35·0
White Coin	1919	64	104	52·5
Ridit	1923	54	92	47·4
Hybrid 128	1924	67	121	48·7
Requa	1931	36	61	46·3
Golden	1947	63	106	49·4
Orfed	1949	82	123	42·6
Elgin	1949	41	56	53·6
Norin 10	1954	55	88	43·0
Brevor	1957	85	134	47·9
Marfed	1959	87	99	60·5
Omar	1959	58	94	49·2
Burt	1959	47	97	39·1
Gaines	1961	40	94	57·8
Daws	1980	37	79	53·4

and amount of carbon released by roots and they need investigation. Such considerations could become important in future plant-breeding programmes.

One of the great excitements in soil microbiology today is the potential for the introduction of the *nif* genes, which are responsible for nitrogenase function, into the plant [9]. Since the first reports of transfer of nif genes between micro-organisms [11,24], the prospect has captured the imagination of scientists and the public. To date, probably the most significant development is that the genes have been transferred to a eukaryotic cell (section 6.2), even though the expression of the function has not been obtained. If this barrier could be overcome, the way might be open to introduce the genes into an organism, such as mycorrhiza, which would infect the plant. Alternatively, the vector might be an organism such as *Agrobacterium tumefaciens* (section 10.4) where the *nif* genes could be incorporated into the plasmid which induces crown-gall tumours; certainly a section of T-DNA on the plasmid can combine with chromosomal DNA in the nucleus of the plant cells [9].

10.10   **Conclusion**

In the past, soil microbiology has gained an image of being a branch of natural history. Whereas it has been important to place the subject on a firm ecological basis, some of the recent developments show its close links with genetics and biochemistry, as well as with the more traditional disciplines related to the study of soils and plants. The potential to manipulate agricultural systems through modification of the soil microflora is now evident. The preparation of rhizobial seed inoculants is already a proven biotechnological application, yet there is scope for more efficient inocula. The value of endomycorrhizal inoculants is not yet proven but if it is, the impetus for achieving the growth of these interesting fungi on culture media in the absence of plants would be much greater.

Unfortunately there have always been openings in this area for 'snake-oil salesmen' who are prepared to sell the farmer miracle preparations containing micro-organisms which, it is claimed, will increase yields dramatically. Although a genuine product could be provided through this route some day, the prospects are not good. Commenting on the value of microbial fertilizers, activators and conditioners it has been concluded [25] that, 'Their mode of action is usually shrouded in mystery and no replicated test data are presented to support claims made for them. Most of the claims are in the form of farmer testimonials. Claims of higher yields or alleviations of some troublesome soil conditions, unfortunately, are often quite appealing to the farmer'. Furthermore, the coefficient of variation in any field trial is often greater than 20 per cent and rarely less than 5 per cent [64]. In screening such products, the claim of the distributor must be examined carefully. Assuming those micro-organisms which are claimed are present, which is often not the case, then glasshouse tests need to be coupled with field trials. If the preparations are claimed to provide nutrients, trials where water is the factor limiting crop growth are pointless. Nutrients may not become available to the crop in the first year, a new equilibrium of nutrients possibly being achieved in two or more years and hence field trials in a single year may show no benefits. This consideration would be particularly relevant if organic methods were introduced into conventional farming systems. The economics of any such introductions would clearly have to satisfy the farmer. Organic methods have not usually been compared with conventional systems where maximum yield potentials are approached and such a comparison is clearly desirable.

By comparison with the miracle products, the preparations that are being developed with a sound scientific basis appear genuine and exciting. These include the plant-growth-promoting rhizobacteria (section 10.4) or seed inoculants, such as *Pseudomonas fluorescens* which produce the antibiotic pyoluteorin against *Pythium ultimum* [33]. However, more research effort will be needed to investigate their action and efficacy at the fundamental and practical level. Such relatively low-cost studies in soil biotechnology should show a good return on investment and help to increase our food productivity even if our net energy resources decline.

It is now about a half century since Selman A. Waksman and others provided soil microbiology with such a sound foundation from which to evolve [65]. It is perhaps ironical that Waksman's own efforts in what we now recognize as biotechnology demonstrated the usefulness of soil microbial activities for antibiotic production in the pharmaceutical industry. Who will now show how other activities can be applied to the benefit of the agrochemical industry and crop productivity generally?

References

1 Abdel Wahab A. M. & Wareing P. F. (1980) Nitrogenase activity associated with the rhizosphere of *Ammophila arenaria* L. and effect of inoculation of seedlings with *Azotobacter*. *New Phytologist*, **84,** 711–21.

2 Allison F. E. (1973) *Soil Organic Matter and its Role in Crop Production*. Elsevier, Amsterdam.

3 Baker K. R. & Cook R. J. (1974) *Biological Control of Plant Pathogens*. W. H. Freeman, San Francisco.

4 Balfour E. B. (1950) *The Living Soil*. Faber & Faber, London.

5 Bingham J. (1981) Breeding wheat for disease resistance. In *Strategies for the Control of Cereal Disease*, eds Jenkyn J. F. & Plumb R. T., pp. 3–14. Blackwell Scientific Publications, Oxford.

6 Brown M. E. (1982) Nitrogen fixation by free-living bacteria associated with plants—fact or fiction? In *Bacteria and Plants*, eds Rhodes-Roberts M. E. & Skinner F. A., pp. 25–41. Academic Press, London.

7 Brown M. E. & Burlingham S. K. (1968) Production of plant growth substances by *Azotobacter chroococcum. Journal of General Microbiology*, **53,** 135–44.

8 Brown M. E., Burlingham S. K. & Jackson R. M. (1964) Studies on *Azotobacter* species in soil. III. Effects of artificial inoculation on crop yields. *Plant and Soil*, **20,** 194–214.

9 Brill W. J. (1981) Agricultural microbiology. *Scientific American*, **245,** 199–215.

10 Buresh R. J., Casselman M. E. & Patrick W. H. (1980) Nitrogen fixation in flooded soil systems, a review. *Advances in Agronomy*, **33,** 149–92.

11 Cannon F. C., Dixon R. A., Postgate J. R. & Primrose S. B. (1974) Chromosomal integration of *Klebsiella* nitrogen-fixing genes in *Escherichia coli. Journal of General Microbiology*, **80,** 227–39.

12 Cochran V. L., Elliott L. F. & Papendick R. I. (1977) The production of phytotoxins from surface crop residues. *Soil Science Society of America Journal*, **41,** 903–8.

13 Chet I., Hadar Y., Elad Y., Katan J. & Henis Y. (1979) Biological control of soil-borne plant pathogens by *Trichoderma harzianum*. In *Soil-borne Plant Pathogens*, eds Schippers B. & Gams W., pp. 585–91. Academic Press, London.

14 Cook R. J. (1977) Management of the associated microbiota. In *Plant Disease*, vol. 1, eds Horsfall J. G. & Cowling E. B., pp. 145–66. Academic Press, New York.

15 Cook R. J. (1980) *Fusarium* foot rot of wheat and its control in the Pacific Northwest. *Plant Disease*, **64,** 1061–6.

16 Cook R. J. (1981) The influence of rotation crops on take-all decline phenomenon. *Phytopathology*, **71,** 189–92.

17 Cook R. J. & Reis E. (1981) Cultural control of soil-borne pathogens of wheat in the Pacific North-west of the USA. In *Strategies for the Control of Cereal Diseases*, eds Jenkyn J. F. & Plumb R. T., pp. 167–77. Blackwell Scientific Publications, Oxford.

18 Cooper R. (1959) Bacterial fertilizers in the Soviet Union. *Soils and Fertilizers*, **22,** 327–33.

19 Council for Agricultural Science and Technology (1980) *Organic and Conventional Farming Compared*. CAST Report no. 84, Ames, Iowa.

20 Dobereiner J. & Campelo A. M. (1971) Non-symbiotic nitrogen-fixing bacteria in tropical soils. *Plant and Soil* special volume, 457–70.

21 Dobereiner J. & Day J. M. (1976) Associative symbioses in tropical grasses. Characterization of micro-organisms and dinitrogen-fixing sites. In *Proceedings of the First International Symposium on Nitrogen Fixation*, vol. 2, eds Newton W. E. & Nyman C. J., pp. 518–38. Washington State University Press, Washington.

22 Dobereiner J., Day J. M. & Dart P. J. (1972) Nitrogenase activity and oxygen sensitivity of the *Paspaulum notatum–Azotobacter paspali* association. *Journal of General Microbiology*, **71,** 103–16.

23 Duff R. B., Webley D. M. & Scott R. O. (1963) Solubilization of minerals and related materials by 2-ketogluconic acid-producing bacteria. *Soil Science*, **95,** 105–14.

24 Dunican L. K. & Tierney A. B. (1974) Genetic transfer of nitrogen fixation from *Rhizobium trifolii* to *Klebsiella aerogenes. Biochemical and Biophysical Research Communications*, **57,** 62–72.

25 Dunigan E. P. (1979) Microbial fertilizers, activators and conditioners: a critical review of their value to agriculture. *Developments in Industrial Microbiology*, **20**, 311–22.

26 Elliott L. F. & Lynch J. M. (1983) Pseudomonads as a factor in the growth of winter wheat (*Triticum aestivum* L.). *Soil Biology and Biochemistry* (in press).

27 Eren J. & Pramer D. (1978) Growth and activity of the nematode-trapping fungus *Arthrobotrys conoides* in soil. In *Microbial Ecology*, eds Loutit M. W. & Miles J. A. R., pp. 121–7. Springer-Verlag, Berlin.

28 Faull J. L. & Campbell R. (1979) Ultrastructure of the interaction between the take-all fungus and antagonistic bacteria. *Canadian Journal of Botany*, **57**, 1800–8.

29 Gilmour J. T. & Gilmour C. M. (1978) Nitrogenase activity of rice plant roots. *Soil Biology and Biochemistry*, **10**, 261–4.

30 Harper S. H. T. & Lynch J. M. (1979) Effects of *Azotobacter chroococcum* on barley seed germination and seedling development. *Journal of General Microbiology*, **112**, 45–51.

31 Harper S. H. T. & Lynch J. M. (1981) The kinetics of straw decomposition in relation to its potential to produce the phytotoxin acetic acid. *Journal of Soil Science*, **32**, 627–37.

32 Howard A. (1940) *An Agricultural Testament*, Oxford University Press, Oxford.

33 Howell C. R. & Stipanovic R. D. (1980) Suppression of *Pythium ultimum*-induced damping-off of cotton seedlings by *Pseudomonas fluorescens* and its antibiotic pyoluteorin. *Phytopathology*, **70**, 712–15.

34 Jensen H. L. (1941) Nitrogen-fixation and cellulose decomposition by soil micro-organisms. III. *Clostridium butyricum* in association with aerobic cellulose decomposers. *Proceedings of the Linnean Society of New South Wales*, **66**, 239–49.

35 Kerr A. (1980) Biological control of crown gall through production of agrocin 84. *Plant Disease*, **64**, 24–30.

36 Kloepper J. W., Leong J., Teintze M. & Scroth M. N. (1980) Enhanced plant growth by siderophores produced by plant-growth-promoting rhizobacteria. *Nature (London)*, **286**, 885–6.

37 Kloepper J. W., Leong J., Teintze M. & Schroth M. N. (1980) *Pseudomonas* siderophores: a mechanism explaining disease-suppressive soils. *Current Microbiology*, **4**, 317–20.

38 Kloepper J. W., Schroth M. N. & Miller T. D. (1980) Effects of rhizosphere colonization by plant-growth-promoting rhizobacteria on potato plant development and yield. *Phytopathology*, **70**, 1078–82.

39 Ko W-H. (1971) Biological control of seedlings root rot of papaya caused by *Phytophthera palmivora*. *Phytopathology*, **61**, 780–2.

40 Lynch J. M., Harper S. H. T. & Sladdin M. (1981) Alleviation by a formulation containing calcium peroxide and lime, of microbial inhibition of cereal seedlings establishment. *Current Microbiology*, **5**, 27–30.

41 Lynch J. M. & Elliott L. F. (1983) Aggregate stabilization of volcanic ash and soil during microbial degradation of straw. *Applied and Environmental Microbiology* (in press).

42 Lynch J. M. & Elliott L. F. (1983) Minimizing the potential phytotoxicity of wheat straw by microbial degradation. *Soil Biology and Biochemistry* (in press).

43 Lynch J. M. & Panting L. M. (1980) Variations in the size of the soil biomass. *Soil Biology and Biochemistry*, **12**, 547–50.

44 Lynch J. M., Slater J. H., Bennett J. A. & Harper S. H. T. (1981) Cellulase activities of some aerobic micro-organisms isolated from soil. *Journal of General Microbiology*, **127**, 231–6.

45 Lynch J. M. & White N. (1977) Effects of some non-pathogenic micro-organisms on the growth of gnotobiotic barley plants. *Plant and Soil*, **47**, 161-70.

46 Martin J. P. (1971) Decomposition and binding action of polysaccharides in soil. *Soil Biology and Biochemistry*, **3**, 33-41.

47 Mehrotra C. L. & Lehri L. K. (1971) Effect of *Azotobacter* inoculation on crop yield. *Journal of the Indian Society of Soil Science*, **19**, 243-8.

48 Metting B. (1981) The systematics and ecology of soil algae. *Botanical Review*, **47**, 195-312.

49 Mishustin E. N. (1970) The importance of non-symbiotic nitrogen-fixing micro-organisms in agriculture. *Plant and Soil*, **32**, 545-54.

50 Mulder E. (ed.) (1979) *Soil Disinfestation*. Elsevier, Amsterdam.

51 Neal J. L., Larson R. I. & Atkinson T. G. (1973) Changes in the rhizosphere populations of selected physiological groups of bacteria related to substitution of specific pairs of chromosomes in spring wheat. *Plant and Soil*, **39**, 209-12.

52 Ohta Y. & Nakayama M. (1971) Utilization of calcium peroxide for agriculture as oxygen supplying material. *Nogyo Oyobi Engei*, **46**, 869-72. (In Japanese.)

53 Papendick R. J. & Cook R. J. (1974) Plant water stress and development of *Fusarium* foot rot in wheat subjected to different cultural practices. *Phytopathology*, **64**, 358-63.

54 Postgate J. R. & Hill S. (1979) Nitrogen fixation. In *Microbial Ecology. A Conceptual Approach*, eds Lynch J. M. & Poole N. J., pp. 191-213. Blackwell Scientific Publications, Oxford.

55 Powell P. E., Cline G. R., Reid C. P. P. & Szaniszlo P. J. (1980) Occurrence of hydroxamate siderophore iron chelators in soils. *Nature (London)*, **287**, 833-4.

56 Rice W. A. & Paul E. A. (1972) The organisms and biological processes involved in asymbiotic nitrogen fixation in waterlogged soil amended with straw. *Canadian Journal of Microbiology*, **18**, 715-23.

57 Ridge E. H. & Rovira A. D. (1968) Microbial inoculation of wheat. *Transactions of the Ninth International Congress of Soil Science, Adelaide*, **3**, 473-81.

58 Shields M. S., Lingg A. J. & Heimsch R. C. (1981) Identification of a *Penicillium urticae* metabolite which inhibits *Beauvaria bassiana*. *Journal of Invertebrate Pathology*, **38**, 374-7.

59 Shipton P. J., Cook R. J. & Sitton J. W. (1973) Occurrence and transfer of a biological factor in soil that suppresses take-all of wheat in eastern Washington. *Phytopathology*, **63**, 511-17.

60 Smith G. E. (1942) Sanborn field: fifty years of field experiments with crop rotations, manure, and fertilizers. *Missouri Agricultural Experiment Station Bulletin*, 458.

61 Stonehouse B. (ed.) (1981) *Biological Husbandry: A Scientific Approach*. Butterworth & Co., London.

62 Tchan Y. T. & Jackson D. L. (1965) Studies of nitrogen-fixing bacteria. IX. Study of inoculation of wheat with *Azotobacter* in laboratory and field experiments. *Proceedings of the Linnean Society of New South Wales*, **90**, 290-8.

63 United States Department of Agriculture (1980) *Report and Recommendations on Organic Farming*. USDA Science and Education Administration, Beltsville.

64 Weaver R. W. (1979) Evaluation of the effectiveness of microbial fertilizers, activators and conditioners. *Developments in Industrial Microbiology*, **20**, 323-8.

65 Waksman S. A. & Starkey R. L. (1931) *The Soil and the Microbe*. John Wiley & Sons, New York.

APPENDIX

The following plant species are referred to only by their common names in the text.

Alfalfa	*Medicago sativa*
Almond	*Prunus amygdalus*
Apple	*Malus pumila*
Asparagus	*Asparagus officinalis* var. *altilis*
Banana	*Musa* spp.
Barley	*Hordeum vulgare*
Bean (French, kidney)	*Phaseolus vulgaris*
Cabbage	*Brassica oleracea* var. *capitata*
Carrot	*Daucus carota*
Cherry	*Prunus avium*
Citrus	*Citrus* spp.
Clover	*Trifolium* spp.
Couch (quack) grass	*Agropyron repens*
Eastern white pine	*Pinus strobus*
Faba bean	*Vicia faba*
Leek	*Allium porrum*
Lettuce	*Lactuca sativa*
Lupin	*Lupinus* spp.
Onion	*Allium cepa*
Oilseed rape	*Brassica napus*
Papaya	*Carica papaya*
Pea	*Pisum sativum*
Peach	*Prunus persica*
Pear	*Pyrus communis*
Plum	*Prunus domestica*
Potato	*Solanum tuberosum*
Red clover	*Trifolium pratense*
Rice	*Oryza sativa*
Rose	*Rosa* spp.
Rye-grass	*Lolium perenne*
Soybean	*Glycine max*
Sugar-beet	*Beta vulgaris* var. *altissima*
Tomato	*Lycopersicon esculentum*
Wheat	*Triticum aestivum*

INDEX

Italicized page numbers indicate that the entry appears in a figure legend.